T0254551

The Rise of Science

Peter Shaver

The Rise of Science

From Prehistory to the Far Future

Peter Shaver
Sydney, Australia

ISBN 978-3-319-91811-2 ISBN 978-3-319-91812-9 (eBook)
https://doi.org/10.1007/978-3-319-91812-9

Library of Congress Control Number: 2018943455

Cover illustration: The iconic "Earthrise" photo taken on 24 December 1968 by the crew of the Apollo 8 spacecraft as they orbited the Moon (seen in the foreground). Photo Credit: NASA. Image Science and Analysis Laboratory, NASA-Johnson Space Center

Printed on acid-free paper

This Springer imprint is published by the registered company Springer International Publishing AG part of Springer Nature.
The registered company address is: Gewerbestrasse 11, 6330 Cham, Switzerland

To Iggy, Max, Lulu and Josie

Preface

A knowledge of the natural and physical world is essential for all of us. Like other animals we are born with senses and instincts, but we have no direct knowledge of the world. It is only when we are born that we can begin to experience and know the world around us. This is urgent and vital for our survival. It is why infants have sometimes been referred to as 'young scientists' and 'learning machines'. They have to learn about the world rapidly in order to be able to cope with it. Curiosity plays a huge role in this process. After several years they have developed intuition and common sense about the world and have learned enough to be able to survive in it.

But some individuals never lose their childhood curiosity and have taken the quest for knowledge far beyond what is required for mere survival. These are the scientists who ask deep questions and systematically explore every niche out of pure curiosity about the world and our place in it. This search for knowledge has gone to extremes over the last few centuries, including the early universe, the atom and the very basis of life itself. The body of scientific knowledge that has been acquired is one of humanity's greatest achievements and a cherished part of our culture. In addition, this knowledge has given us great power, leading to the high living standards that we enjoy today.

The story of how science originated and developed over the centuries is a fascinating one. Its origin was perhaps improbable, and it actually disappeared a few times over its history, but in the last few centuries its growth has been spectacular and exponential. This book follows this remarkable story, explores many intriguing aspects of science, considers its place in modern society and reflects on its future and how it may further change the world.

I am grateful to many people for their encouragement, advice and knowledge provided over the four years I took to write this book. I would especially like to thank Ron Ekers (CSIRO Astronomy & Space Science, Sydney), Ryszard Maleska (Research School of Biology, Australian National University, Canberra) and Steve Simpson (Charles Perkins Centre, University of Sydney) for having taken the time for lengthy discussions with me on several of the topics in this book and for their opinions and other information. As always, my wife Jenefer read over drafts of the book, and we had many wonderful conversations about the subjects of the day over dinner.

Several colleagues and friends generously took the time to comment on advanced drafts of the book. In particular, Gordon Robertson (University of Sydney) read through a draft of the book thoroughly, making copious comments and suggestions and raising several thought-provoking issues. Richard Schilizzi (University of Manchester), Elizabeth Jeffreys (Exeter College, Oxford), Dominic O'Meara (University of Freiburg) and Anne Tihon (Université Catholique de Louvain) provided helpful input concerning the demise of Greek and Byzantine natural philosophy. Martin Harwit (Cornell University), Ken Kellermann (U.S. National Radio Astronomy Observatory), Ted Brown (Visiting Professor at the University of Sydney and Director (Emeritus), NYS Institute for Basic Research), Neville Chalkley, Russell Stewart and David Woodruff also gave helpful comments, and I would like to thank them all for their input.

Access to the latest scientific publications in all fields was facilitated through an Honorary Associateship kindly granted to me by the School of Physics, Faculty of Science, the University of Sydney.

My editor at Springer, Ramon Khanna, was very supportive and gave me several helpful suggestions to improve and balance the book. I am most grateful both for his generous input and for his guidance throughout the publishing process.

I must certainly acknowledge that the Internet has been essential in putting this book together—countless websites covering almost all conceivable topics for inspiration, information and cross-checking. To say that these have 'supplemented' my very substantial library of books and discussions with colleagues would be a gross understatement. I don't think I could have written this book 20 years ago.

Sydney, Australia
June 2018

Peter Shaver

Contents

1

Introduction

Modern science and technology have given us knowledge and living standards far beyond the wildest expectations of our ancestors who lived only a few lifetimes ago.

Just a few lifetimes ago (using as a timescale our modern lifespan of 80 years, a tiny fraction of our existence as a species), our ancestors had no idea what matter is made of, how life works, or what lies beyond our solar system and the nearest stars. Electricity was just a curiosity and the atom was just speculation.

Now we know about the atom and its constituents, we understand the basis of life, and we have explored the universe. Today electricity is widely distributed over countries and powers our industries and homes, and our ancestors would be even more amazed by our use of atomic power. They would have looked up at the Moon and wondered what it is made of. Now we have walked on the Moon and sent spacecraft to all the planets in the solar system.

Three lifetimes ago our ancestors travelled by foot or on carts over dirt lanes; the wealthy few travelled by horse or horse-drawn carriages. Today we routinely travel by car at 100–200 km per hour over paved roads and superhighways.

Back then travel overseas was in wooden boats with sails, taking months, and long distance communication was by letters carried by horse or sent overseas, again in wooden boats with sails. Today we travel halfway around the world in just hours in huge jet airplanes that fly 10 km above the ground and close to the speed of sound, and we have virtually instantaneous communication worldwide by telephone, email and social media.

Our ancestors at that time lived in cottages, wooden farmhouses or stone houses in villages and towns, with only fireplaces for warmth and candles for

P. Shaver, *The Rise of Science*, https://doi.org/10.1007/978-3-319-91812-9_1

light. The water for household use had to be carried in by hand. Today we live in well-insulated homes with thermopane windows, central heating, air conditioning and electric lighting, and we take for granted the indoor hot and cold running water that is always instantly available.

Just a few lifetimes ago men worked the farms manually, with help from horses or oxen, and women washed the clothes by hand. Today only a small fraction of people work on farms, with machines doing the heavy work, and the washing machine has helped to liberate women.

At that time the food was locally produced and had to be rapidly consumed, preserved or put on ice before it decayed. Today we have foods from around the world always available in supermarkets and refrigerators at home to keep it fresh.

Back then the very lives of our ancestors were at the whim of the weather. Now we have weather satellites viewing the entire globe and massive computer simulations to warn us of imminent disasters and make predictions of remarkable accuracy.

Simple music and skits at that time were produced in the home or village by family members, locals or small traveling groups; only the wealthy could experience chamber music, concerts, operas and stage plays. Today we have all of the world's music and a huge variety of entertainment by the world's greatest performers available wherever we are at the touch of a button on our iPods, smartphones, CD players, television sets and computers with streaming devices connected to the Internet.

Three lifetimes ago the only pictures were drawings and paintings. Now we take trillions of digital photographs every year.

Life expectancy back then was only 30 years or so. Today it is over 80 in much of the developed world. At that time bloodletting was a common remedy, and the causes of diseases were totally unknown. Now we understand and have cured many of them, we have powerful imaging devices to see inside the human body and sophisticated drugs and methods of diagnosis and operations, saving millions of lives.

The world a few lifetimes ago was almost totally dark at night, and our ancestors would be astonished by the dazzling sight of modern New York, Hong Kong or Tokyo at night.

Our ancestors just a few lifetimes ago would be left speechless by the sight of the Space Shuttle taking off, or by the notion of a submarine that can reach the deepest parts of the ocean, and they would consider a modern computer or smartphone to be a miracle.

The science fiction writer Arthur C. Clarke once remarked that any sufficiently advanced technology, in the eyes of a much less developed society, would be "indistinguishable from magic".

Our knowledge of the world has increased exponentially, and no one individual can possibly keep up. There are experts in every niche. But thanks to the Internet most of the world's knowledge is now readily accessible to all of us.

Throughout almost all of the existence of humanity our ancestors had no expectation of progress; they eked out an existence and hoped only that tomorrow would be like today. But now we live in a world of phenomenal exponential growth; we are addicted to progress and we take it for granted as we are so accustomed to it. Children born today know nothing but our modern world of smartphones, cars and airplanes.

The past two or three lifetimes have been truly exceptional. The world a few lifetimes ago would have looked much as it did hundreds or even thousands of years before. It is only since then that our lives have changed so dramatically, thanks to science and technology.

But the story of science goes back much further, to the dawn of humanity.

2

A Brief History

2.1 Out of the Mists of Time

Our ancestors emerged out of the mists of time with a developing conscious awareness of themselves and the world they found themselves in.

Some 7 million years ago they parted ways with the chimpanzees. At that time our ancestors and the chimpanzees started to evolve separately, and eventually became entirely different species. The chimpanzees stayed in their natural habitat, but our ancestors had a very harsh and challenging experience. They moved away from their familiar forests into entirely new habitats, starting with the open savannahs of Africa, and eventually the cold plains of Asia, the mountains and ice of Europe, and ultimately the rest of the world. These new environments would have forced significant evolutionary change on our ancestors as they struggled to adapt. The result (so far) is us.

One of the first crucial evolutionary adaptations was bipedalism, the ability to walk on two feet. It required a number of significant anatomical changes in the spine, pelvis, legs and feet. But it brought major advantages. It became possible to see other large animals—both predators and prey—at much greater distances over the grassy savannahs. It was possible to move faster and more efficiently. The hands were freed for other tasks, something that was to become ever more important with time. The footprints of our early ancestors have been found in solidified volcanic ash dated at over 3.5 million years ago.

As our early ancestors (like us) had no natural offensive or defensive weapons or armour (such as the antlers, large teeth, claws, and shells of other animals), the development of tools was a very important step. Without

P. Shaver, *The Rise of Science*, https://doi.org/10.1007/978-3-319-91812-9_2

them, our ancestors had just their bare hands. The oldest known stone tools date back 2.5 million years. Initially they were no more than suitably-shaped stones that could be used for scraping the flesh of dead animals. But recognizing their usefulness was a small but very significant step. The tools developed slowly over the ages.

The first use of fire took place over a million years ago. It would probably have been in the form of a burning branch removed from a wildfire caused by a lightning strike, and taken to the camp. The resulting campfire would have provided warmth, some protection from wild animals, and a means of cooking food (which may have been particularly advantageous for energy-hungry brains). It could be sustained by adding more branches as required. The big breakthrough would have been the discovery of a way to make fire themselves, using friction. Then they had a portable tool of enormous value.

Another major evolutionary change was the increase in the size of the brains of our ancestors. A few million years ago they were no larger than those of chimpanzees today, about 300–400 cm^3. But then they began to increase with time, and today they are about 1400 cm^3 in volume. The prefrontal cortex, the seat of our highest cognitive powers, is seven times bigger than that of chimpanzees. Considering the huge cost of such large brains—the difficulty in giving birth and the energy requirements of the brain (about 50% of the total intake in the case of infants)—the benefits must have been huge. The reason for this extraordinary evolution is not yet clear, but the advantages of large brainpower today are obvious to all of us.

A knowledge of materials for tools, the nature of fire, which berries are safe to eat and which are poisonous, the behaviour and migration patterns of animals and other crucial facts about the world was vital for survival, and would have been passed down over the ages from generation to generation over millions of years. This was science at its most elementary level. Over most of that time there was no spoken or written communication—the young would have learned from observing the practices of the old. But a more general form of communication would clearly have been highly beneficial.

So the origin of language would have been a monumental step forward. When and how did it happen? Unfortunately, because of the obvious lack of direct evidence it is very difficult to answer these questions, although there are many hypotheses. Language and speech may be considered to be different things, but the development of the larynx would certainly have been an important step. The combination of a large brain, a fully developed larynx and language would have given our ancestors an enormous advantage.

The evolution of our early ancestors did not follow just one single lineage. Over time it led to as many as 27 different lineages according to recent estimates, and a tentative 'family tree' has been drawn. There may have been several different species living on the planet at any one time. Eventually all of them became extinct, except us. *Homo* is the genus of primates which includes modern humans. *Homo habilis* appeared in the fossil record about 2.8 million years ago, and probably made the first tools. *Homo erectus* and *Homo ergaster* were probably the first to use fire and complex tools, and were the first to leave Africa and live in Europe and Asia some 1–2 million years ago. *Homo neanderthalensis* (the Neanderthals), with brains as large as ours, were living in Europe as early as 400,000 years ago. Anatomically modern humans (*Homo sapiens*), our direct ancestors, first appeared about 200,000 years ago, and there was a major wave of migration out of Africa some 50,000 years ago. There are indications from genetic analysis that humans may have suffered a 'population bottleneck' along the way, reducing the population to just a few thousand and dangerously restricting the genetic diversity, but obviously we recovered. The Neanderthals disappeared from the fossil record about 30,000 years ago. Ours is the only surviving species of the genus *Homo*.

Over most of the last 2 million years the technology of tools evolved only very gradually. But then, about 50–100 thousand years ago, innovations of several kinds began to appear. Some researchers see evidence for a 'Great Leap Forward' in technology and culture about 50 thousand years ago, while others see a more gradual development over the last 100 thousand years. In any case, significant changes were taking place. The evidence includes the use of more effective tools made of bone, composite weapons such as spears with stone or bone heads, the bow and arrow, fishing, cave paintings, jewellery, transport of materials over long distances, and burial rites. Most significant were the new activities of a symbolic and abstract nature. Anatomically modern humans were becoming behaviourally modern humans.

Religion gradually developed. All the known cultures of the world practiced some form of religion. Then as now it involved a belief in supernatural entities, and concerned itself with the spiritual aspect of the human condition, life and death, and the possibility of an afterlife. Each community created gods in its own image. They were anthropomorphic, and mirrored the values of the community closely. They helped to make sense of the human condition. There were gods for everything, and it is understandable that primitive people would have invented gods, spirits and myths to explain terrifying or mysterious natural phenomena such as lightning, thunder, comets, meteors, eclipses, rainbows, earthquakes, birth and death. Of course, these natural phenomena are all explained by science today, but in early times it would have been comforting to 'understand' them in terms of gods, and it would certainly have been reassuring to believe in an afterlife. Religion was also a powerful unifying force amongst the members of a community.

About 10,000 years ago some of our ancestors gave up their hunter-gatherer lifestyle. They settled in one place, started agriculture, and domesticated animals. These developments undoubtedly took place over an extended period of time. They learned how to grow a few crops by first observing that they originated in seeds, and gradually improved their techniques. The wolves that followed them around from campsite to campsite eating leftover scraps in earlier times eventually became tame dogs. Our ancestors learned to keep other animals which could be used for milk, food and clothing, such as cows and sheep. Not all of our ancestors gave up the hunter-gatherer lifestyle, and some have continued that lifestyle up to the present day in remote locations as diverse as New Guinea, central and southern Africa, the Amazon basin, Siberia, Alaska, northern Canada, Western Australia, Tierra del Fuego, Malaysia and the Andaman Islands. But from those that chose a settled existence there arose villages, towns, and eventually entire civilizations.

2.2 The Early Civilizations

The major early civilizations (Mesopotamia, Egypt, Indus River, Yellow River, Mesoamerica and South America) all emerged between about 5000 and 3000 years ago, a small range compared to the time since the exodus out of Africa some 50,000 years ago. It is particularly interesting that at least those in Eurasia and America must have emerged independently of each other. This (like the advent of the early religions and writing) seems to suggest that all humans were developing almost in step with each other, even though they may have been on opposite sides of the planet.

These civilizations typically had large population densities, cities, centralized authority, specialization of labour, social stratification, taxation, monumental building, dependence on intensive agriculture and water management, surplus production and writing. Four of them were located in close proximity to large sources of water: the Tigris and Euphrates rivers (Mesopotamia), the Nile river (Egypt), the Indus river (Indus River Valley), and the Yellow River (China). The others, located in Mesoamerica and South America, relied on large irrigation networks.

Irrigation was essential to all of them. A range of major works and innovations took place. They built canals, dikes and dams. Devices based on the principle of the lever were used in Egypt and Mesopotamia to raise water from rivers and pour it into canals. The Mesopotamians developed pumps consisting of series of buckets attached to ropes to draw water up from wells. As part of these systems they invented the pulley.

Oxen and horses were domesticated. To work the land, the oxen were used to pull ploughs. A 'seeder plough' was developed—a plough with a funnel attached to drop seeds into the furrow.

Systems of roads were built to unify the far-flung empires. The wheel was initially invented about 6000 years ago to facilitate the shaping of pottery, but the major breakthrough for transportation was the invention of the axle, so that two wheels attached to the axle could rotate in an upright position over the ground. Carts and chariots were the result. Egypt had its own freeway running the entire length of the country—the Nile. The Egyptians invented sailboats to ply the Nile, coasting northwards with the current and sailing southwards with the wind. Trade would have been an important ingredient of these economies.

Monumental structures were hallmarks of these civilizations. Egypt's Great Pyramid is the largest solid stone structure every built. The Great Wall of China, completed in 221–207 BC and 22,000 km in length, is the largest building project in history. The building of such vast structures required planning, design, massive manpower, and technical expertise. Complex buildings also required innovations; the stone arch was a very important contribution from the Mesopotamians.

Metallurgy and chemistry were important in several of these early civilizations. The metals included copper, bronze, tin, silver and gold; the metalworking involved some complicated technologies, from mining and smelting to hammering or casting the metal into useful objects. There was also great interest in alchemy.

To preserve the bodies of the dead, the Egyptians developed mummification, which required a knowledge of the chemical compounds that had preservative properties. The Mesopotamians learned the properties of various chemical substances, and applied them to the manufacture of a number of products, from soap to clothing, glass and glazed pottery.

Medicine was widely practiced. The early Egyptians had a strong tradition of medical science, as recorded on papyrus textbooks. They also developed pharmaceuticals in the form of natural extracts; the curative powers of several of them have been confirmed by modern science.

Some form of writing was one of the defining characteristics of the early civilizations, and different techniques were developed. The earliest was the cuneiform system of writing on clay tablets, originating in Mesopotamia about 3000 BC. The hieroglyphs emerged in Egypt at about the same time. As in other cultures, both of these eventually evolved to represent sounds of the spoken language.

The early civilizations developed systems of arithmetic and geometry. These included linear, quadratic and cubic equations, methods to compute areas, volumes and compound interest, and approximate determinations of the value of π. They were obviously of importance for the agriculture and engineering of the time, for the accounting required by a large bureaucracy, and for trade. Standardized weights and measures were developed, and the first money was introduced.

All these civilizations conducted astronomical observations. These were of importance for a variety of purposes, ranging from religion and astrology to predictions of major astronomical and terrestrial events. The highly regular annual flooding of the Nile could be predicted in advance from the first appearance of the star Sirius above the horizon. Mesopotamian astronomers could predict the paths of the principal heavenly bodies and the time of appearance of a new Moon. They connected the stellar dots to name the constellations, divided the sky into the signs of the zodiac, and identified the Milky Way. Their detailed records of the movements of the stars and planets, the eclipses of the Sun and Moon, and the appearances of comets provided a substantial database for the Greeks who followed. They devised the 360-degree circle and the sixty-minute hour.

So elementary science and the resulting technologies were clearly significant in these early civilizations. Many of the impressive innovations and developments were done for purely utilitarian purposes—to serve the state and its objectives. Higher learning with practical applications was supported by the state and religious leaders. Knowledge was certainly useful for agricultural management, state record-keeping and administration, trade and commerce, architecture and engineering, medicine, religion, the calendar, and astrology. It was maintained by legions of anonymous scribes and bureaucrats, employed by the state for its purposes.

But none of these huge civilizations produced 'natural philosophy', the rational study of the intrinsic properties and workings of the natural and physical world—the basis of modern science. Not a single natural philosopher is known to us from any of these civilizations, in spite of their wealth, size and millennia of existence. Why?

The poor peasants, villagers and workers would not have had the education, leisure or freedom to contemplate the world, and they would have been immersed in the tentacles of small village life. Their focus was mere survival. Any learned individuals would have been employed and enmeshed in the workings of the state, and would probably not have had the inclination, time or permission to look beyond its purposes. And the huge bureaucratic civilizations would have been stultifying and uncreative.

But the most daunting impediment was undoubtedly religion. These early civilizations were awash with gods, spirits, demons and endless myths. The worldviews of the time were completely religious and mythological. The well-established religions explained everything. It was customary in those times to answer a question by invoking a god or inventing some other mythological explanation. There were hundreds, if not thousands of gods, and many more spirits and myths. Religions were fully integrated into daily lives and rituals. In their various forms they looked after the needs of individuals, and provided a belief in a higher power and life after death. Religion was deeply embedded and part of the social fabric of society.

In the face of these all-encompassing religions and mythologies there was no need and no room for a totally different view of the world. The religions already explained everything about the world, and it was inconceivable that mere mortals could find any truths about the world that were not already known to these religions. The very existence of these religions would have been a major obstacle to independent thought, and the priesthoods would certainly have suppressed any attempt at free thinking. And the truths of the universe may have been considered absolute and god-given, in which case it was not man's place to question them or suggest an alternative. The vengeful wrath of the gods would in itself have been enough to deter free thinking.

And then there was the marriage of religion, politics and power. Ancient Egypt was a theocracy. The Pharaoh ruled by a mandate from the gods. He represented the will of the gods through laws and policies. Under him was a state hierarchy, bureaucracy and military, and a powerful priesthood that helped to maintain the status quo. The entrenched privileges of those in power in such a monolithic hierarchical civilization would have been enormous, and it would obviously have been a daunting prospect to propose a totally different worldview that might threaten them.

So it is perhaps not surprising that these early civilizations did not produce anything approaching natural philosophy. That had to await the great Greek thinkers.

2.3 The 'Greek Miracle'

The first of the two fundamental steps in the Rise of Science occurred in the sixth century BC, in ancient Greece.

Individual thinkers began to contemplate the world and their place in it. Instead of mindlessly obeying a priestly elite or explaining things away by inventing yet another mythical god, they tried to make sense of everything based purely on observation, rational thought and causes that were part of the natural world. They thought that there was intrinsic order behind the phenomena of nature, which could be discerned by the human mind. Religion played no role in this process. The Greek thinkers sought the fundamental properties and the underlying principles of the universe.

This 'natural philosophy' is far deeper, broader and more profound than what we normally think of as science. It explored the natural causes of events. It sought the fundamentals that govern the world. It considered the very nature of reality. It questioned the origin and evolution of the universe. It speculated on the elements and atoms that may comprise matter. It asked questions about life and death, the basic principles of physics, the shape and size of the Earth, the nature of the solar system, and the anatomy of animals and humans. It explored the fields of biology, botany, zoology, geology and psychology, and it questioned the possibility of souls and the role of the brain. It raised questions about the natural world without limit. It pursued knowledge for its own sake.

The worldview changed from one of religion and mythology to one of causes that were part of the real world itself, and accessible to rational thought. The world was to be explained not by religion but by science. It was a monumental achievement—a scientific worldview.

It is very difficult for us to realize how absolutely radical and revolutionary this was, as we are so used to thinking that way now. It was an astonishing breakthrough, and it happened only once over the entire history of the world.

Why Greece? Why was it the Greeks who made this historic breakthrough? A major reason was undoubtedly the fragmentary nature of the civilization they lived in, due in turn to the complex geography of Greece, subdivided as it is by mountains, valleys, hills, rivers and waterways. Ancient Greece was comprised of hundreds of decentralized individual city-states, within each of which power was shared by a small group of aristocrats or merchants. And freedom was essential. Citizen in those cities were free from the confines of the village, from the state, and from religion.

Ancient Greece was perhaps unique in that, although religions, cults and myths certainly existed, they were fragmented, often with different gods in different cities, and some individuals even had their own personal gods. This all seemed suspiciously arbitrary. And some of the gods were capricious, which would not exactly have pleased everyone. And finally, there was no overall priestly caste to impose dogma, so even religion itself could be questioned. The door was open for alternative views.

If some of these citizens had independent wealth, they could take the time to imagine and think radically new thoughts. Most of them knew how to read and write, they were relatively well educated, and had some knowledge of the wider world through traveling. The Greek language was innovative and relatively easy to read and write, and this certainly would have been conducive to the rise of sophisticated philosophy. Democracy itself was a Greek innovation; citizens could revel in free debate and novel ideas of any kind could flourish.

With a long coastline and many islands Greece was a seafaring nation, well connected to the wider world through trade and commerce, and exposed to a wide diversity of cultures and ideas. Most of the first natural philosophers lived in Ionia, which was the wealthiest and most urbanized part of Greece at that time. Small informal 'schools' of natural philosophy developed around the more famous thinkers. In all there were hundreds, perhaps thousands, of these Greek thinkers. Some of the most famous are mentioned below.

Thales of Miletus (c. 625–545 BC) was the first of these philosophers. Miletus was a bustling city on the coast of Asia Minor, and as a trade centre it would have been a rich crossroads of ideas and influences. Thales was not only a philosopher. He was worldly-wise and a polymath. According to various sources he was born into a wealthy family, was well educated, and became involved in business and politics. He probably made a trip to Egypt, which would have given him valuable experience. He provided crucial advice in the defence of the Ionians against the Persians, which brought him fame. He is said to have predicted the solar eclipse of 28 May 585 BC (an astonishing achievement at the time, if true), which abruptly ended a war. He was a 'man of the world'. In the midst of all this he turned his talents to natural philosophy.

Thales sought to explain the workings of the universe in natural rather than supernatural terms. His rejection of mythological explanations was the key innovation that was followed by subsequent Greek philosophers over the next thousand years. He is therefore widely considered to be the 'founding father of natural philosophy'. Like many of the philosophers to follow, he covered a host of topics. He is considered to have been the first true astronomer and mathematician, and he also worked on metaphysics, ethics, history, engineering and geography. He calculated the height of a pyramid from triangulation, he measured the distances of boats from a harbour, and he explored magnetism and static electricity. He hypothesized that all of nature is based on a single substance: water. He had two disciples, Anaximander (c. 610–546) and Anaximenes (c. 585–528), and they all freely engaged in critical discussions of the theories of the others. Thales had a profound influence on the other Greek thinkers to follow, and therefore on the development of Western philosophy.

Pythagoras of Samos (c. 570–495 BC) was an Ionian Greek philosopher and mathematician, best known for the Pythagorean theorem. He also contributed to music, astronomy and medicine. He experimented with the sounds from different lengths of string, and is said to have discovered that the pitch of the sound is inversely proportional to the length of the string. This discovery led him to speculate that all physical phenomena may be understood by fundamental mathematical relationships, and his search for such relationships led him to the Pythagorean theorem. Regressing a bit into mysticism, he founded a cult based on a belief in cosmically significant numbers. They believed that the movements of the heavenly bodies produce a 'music of the spheres' in space. He considered the Earth itself to be a sphere. Many discoveries were attributed to him in a variety of fields.

According to Empedocles (c. 500–430 BC), the components of the world are earth (i.e. solid), water (liquid), air (gas) and fire (heat), separately or in various mixtures (the four 'ancient elements'). Leucippus (fl. fifth century) was the originator of the atomic theory, and he asserted that everything in the material world has a natural explanation. His most famous student was Democritus.

Democritus (fl. 420 BC) is widely considered to be the father of modern science. Following from Leucippus, he elaborated on a formulation of an atomic theory. He believed that all matter is composed of atoms (from the Greek word *atomos*, meaning indivisible), clustered in groups. Different materials are comprised of different types of atoms. Empirical arguments were given for their theory. As matter can change shape, the atoms must be separated by empty space. Democritus suggested that light consists of atoms in transit. He speculated on subjects as diverse as the lives of the first humans and the contents of the universe. He held that the universe was originally composed of atoms in chaos, collisions ultimately forming larger units such as the Earth. He maintained that every world has a beginning and an end. Some of these concepts may seem rather familiar to our modern ears. He also wrote on epistemology, aesthetics, mathematics, ethics, politics and biology.

Considered to be the father of medicine, and composer of the famous Hippocratic Oath, Hippocrates of Cos (c. 460–370 BC) directed the most famous school for physicians of the time. In keeping with the philosophical traditions of his predecessors he held that diseases originated from natural (not supernatural) causes, and had to be cured by physical remedies. He believed that sickness results from an imbalance amongst the body's four main fluids, or 'humours'. Others over the following two centuries may actually have written most of the works attributed to him, but the Hippocratic Corpus has had a huge influence, and many medical textbooks survived from that era.

Although Socrates (c. 470–399 BC) was a towering figure in philosophy, he concentrated on moral issues and did not make a direct contribution to science. But his method of inquiry, known as the Socratic method, did have an influence on the scientific method. It involved arguments that can lead to disproof or refutation of a given thesis; knowledge may be possible, but it has to be able to stand up to scrutiny. He never wrote anything himself, and all information about him came from others, in particular his student Plato.

Plato (c. 428–348 BC) was a central figure in the history of Western philosophy. His teacher was Socrates, and his most famous student was Aristotle. The three of them are said to have laid the foundations of Western philosophy and science. Plato's interests included epistemology, metaphysics, mathematics, logic, rhetoric, ethics, justice, politics and education. He held that reality as we experience it is a mere replica of ideal intellectual forms. In his Allegory of the Cave, he described prisoners finally emerging to see the higher truth. The real world, he said, can only be deduced through rational thought. Mathematics was central for Plato; mathematical truths are perfect, unlike the physical world of our senses. He added to Empedocles' four elements a fifth which he called 'aether'—the pure substance that fills the upper regions of the cosmos.

Of all the Greek philosophers, Aristotle (384–322 BC) had by far the greatest impact on the development of science. He studied and wrote on many subjects, including metaphysics, physics, geology, biology, zoology, medicine, anatomy, physiology, psychology, ethics, linguistics, logic, rhetoric, poetry, music, politics and government. His vast work on the natural sciences (some 170 works) had a dominant influence through the Islamic, medieval and Renaissance times right up to the Scientific Revolution in the seventeenth century.

Aristotle developed what at that time seemed a common-sense view of the world. He started with Empedocles' four fundamental elements (earth, water, air and fire), and, like Plato, added a fifth element, the aether (or 'quintessence'). He held that the Earth is a sphere; his evidence included the fact that the Earth's shadow on the Moon during an eclipse is a circle, and the fact that the hull of a departing ship disappears below the horizon before the sails do. As things fall towards the Earth, the Earth is naturally at the centre of the universe. Water naturally tends towards a sphere surrounding the Earth, and air tends towards a higher sphere. Fire naturally tends upwards. Aristotle adopted the notion that the heavenly bodies are attached to a series of concentric spheres as proposed by some other Greek philosophers. These are made of the fifth element, quintessence. He held that the universe is a vast bounded sphere with no beginning and no end in time.

He made observations covering an extraordinarily wide range of natural phenomena, and systematized what was known at that time. He discussed the effects of force and inertia. He observed the properties of the lever. He maintained that a heavier object falls faster than a lighter object. He worked on the field of optics. He became the 'father of zoology', classifying hundreds of species. He believed that there is no sharp boundary between living and non-living things, and that there is a continuum from plants and animals to humans. He regarded the heart as the seat of intelligence, and he suggested the possibility that the soul may exist apart from the body.

Aristotle distinguished between 'natural motion' and 'violent motion'. In natural motion, the five elements mentioned above tend towards their natural places (e.g. an object falling to the Earth, or a bubble of air rising in water). There is no force to be explained. In violent motion, on the other hand, the natural tendency of an object is violated (e.g. a thrown object), and a force must be invoked to explain this unnatural change. He studied many kinds of change, and proposed four kinds of causes: the material cause, the formal cause, the efficient cause, and the final cause. Of these, the efficient cause is most similar to what we mean by 'cause' today.

The most intriguing of these is the final cause—the goal or purpose of the change. Such a notion is called 'teleology', and is not part of modern science. Given that humans are used to doing things for a purpose, it is natural that they might ascribe 'purpose' to events in the natural world. And, of course, it fits in with many religions that believe in a purposeful god. But it is totally incompatible with the proven principles of modern science, in which events follow from past causes, not towards future goals; they result from nature itself, not some underlying purpose.

Aristotle's hypotheses were also at odds with modern science because they were qualitative rather than quantitative, they made no predictions, and no experiments were made to test hypotheses against the real world.

Nevertheless his work, because it was so vast and comprehensive, and because it seemed to be compatible with so many common-sense views, was widely accepted as the received wisdom for almost two thousand years. After some modifications in the thirteenth century, it became essentially the official philosophy of the Roman Catholic Church. As a result it was initially an impediment for the new concepts of science that were introduced in the sixteenth and seventeenth centuries, as we shall see.

Greek philosophy received a boost with the construction of the Great Library of Alexandria in the third century BC. Its ambitious goal was to house all the knowledge of the world; at its height it contained hundreds of thousands of scrolls. It was a major centre of scholarship for hundreds of years, and many notable thinkers worked there.

Following from Democritus, Epicurus (341–270 BC) established a system which saw the workings of nature in terms of atoms which randomly collide. Upon death, the dispersal of the soul's atoms preclude the possibility of an afterlife. The aim of life is the pleasure to be found in moderation and contemplation. Epicurus was an early proponent of the scientific method, insisting that nothing can be believed without being tested by direct observation and logical deduction.

In his famous work *The Elements*, Euclid (fl. 300 BC) synthesized the mathematical knowledge of the Greeks up to his time. It served as a major textbook until just a century ago. He deduced the principles of 'Euclidean geometry' from a small number of axioms. He applied his expertise to fields such as optics, harmonics, and astronomy.

Herophilus (c. 330–260 BC), considered to have been the first anatomist, performed dissections of human cadavers and vivisections of living animals to understand the workings of the body. He realized that the brain is the centre of the nervous system, and identified some of its regions. His contemporary Erasistratus (c. 315–240 BC) recognized the heart's main function as a pump, and discovered that it contains four major one-way valves.

In contrast to the views of Aristotle and most other Greek thinkers, Aristarchus of Samos (c. 310–230 BC) proposed that the Sun, rather than the Earth, is at the centre of the universe. He estimated the relative distances of the Sun and the Moon from the Earth, and their sizes. His heliocentric model was rejected by others for common-sense reasons: a spinning Earth racing around the Sun would cause high winds, objects would be flung off its surface, and bodies thrown upwards would land somewhere else. A more technical issue was that of 'stellar parallax': if the Earth orbits annually about the Sun, then the observed relative positions of the stars would change over a 6-month period, something that was not observed. Thus, it was not Aristarchus' heliocentric model, but rather the classical geocentric model that finally emerged from the Greek times down through the ages.

The greatest mathematician of the Greek philosophers was Archimedes (c 287–212 BC). In addition he was an outstanding inventor, and he also made contributions to physics and astronomy. He developed ways to determine the areas and volumes of a variety of two and three dimensional surfaces, a novel way to calculate the value of π, and he anticipated modern calculus by using the concept of infinitesimals. He became famous for discovering Archimedes' Principle: a body immersed in a fluid experiences a buoyant force equal to the weight of the fluid it displaces. He is said to have discovered the principle while taking a bath, and famously exclaimed "*Eureka!*" (I have found it!). He produced a formula that explains how levers work. He invented the Archimedes' screw for pumping water, devised engines of war, produced a working model of the solar system, a celestial globe and other wonders.

An ingenious measurement of the circumference of the Earth was made by Eratosthenes (c. 284–192 BC). He knew that at noon on the summer solstice in the Egyptian city of Syene the Sun would be directly overhead, as its rays then reach to the bottom of a deep well located there. So at that exact time he measured the angle of elevation of the Sun in Alexandria, almost due north of Syene. Knowing the distance between the two cities, it was an easy matter to compute the circumference of the Earth, assuming it is a sphere.

Hipparchus (c. 190–120 BC) was the ancient world's greatest astronomer. He catalogued 850 stars with their positions, discovered the 'precession of the equinoxes' (slight apparent motions of the stars now known to be due to the wobbling of the rotation axis of the Earth itself), calculated the distance from the Earth to the Moon, and invented a number of new astronomical instruments.

An extraordinary example of the technology possible at that time has been recovered from a shipwreck off the Greek island of Antikythera. It is called the Antikythera mechanism, and is thought to have been built in the second century BC. It contained more than 30 gears, and predicted astronomical positions, the risings and settings of major stars and constellations, the phases of the Moon, the movements of the planets, eclipses and more. It was certainly the most complex device known from ancient Greece, and is considered to be the first known 'computer'. It has been suggested that Hipparchus, and perhaps also Archimedes, might have worked on it. This sophisticated technology remained unsurpassed for the next two millennia.

Ptolemy (c. 100–178 AD) is famous for his model of the motions of the heavenly bodies in which the Earth is at the centre of the universe, published in a treatise known as the *Almagest*. His model was presented in tables which could be used to determine the past or future positions of the heavenly bodies. In his model, the Earth is a sphere and does not move. It is at the centre of the cosmos. The heavenly bodies are perfect spheres that move about the Earth in circles; in order of distance from the Earth they are: the Moon, Mercury, Venus, the Sun, Mars, Jupiter, Saturn and finally the sphere of the fixed stars. To account for some peculiarities in the observed motions of the planets ('retrograde motion' is the apparent backwards motion sometimes seen), his model employed devices called 'deferents', 'epicycles' and 'equants'. The *Almagest* also gave a list of 48 constellations, whose names have survived to the present, containing 1022 stars. It superseded most previous Greek works on astronomy, and became widely accepted. It became one of the most influential scientific texts of all time, and remained the dominant astronomical model for predicting the motions of the heavenly bodies over the next 1400 years.

In another monumental treatise, *The Geography*, Ptolemy presented the entire geographical knowledge of his time. *The Geography* contained the geographical coordinates of all the world known to the Roman Empire at that time, along with maps and discussions of methods and data.

The physician Galen (c. 129–210 AD) became famous for his work on human anatomy. He proved that both arteries and veins transport blood. He wrote hundreds of works on medical science; he conveyed much of the Hippocratic tradition, but he also went much further, adding anatomical and physiological aspects. He had a dominating and lasting influence on Western thought up to the Scientific Revolution in the seventeenth century and even beyond.

The period of the great philosophers of Greece had lasted for over a thousand years, from Thales to Galen and a few later thinkers, peaking at about 500–300 BC, the period of Socrates, Plato and Aristotle. It was an extraordinary time of free thinking and speculation about the world we live in. But it eventually and gradually faded away, as shown in Fig. 2.1.

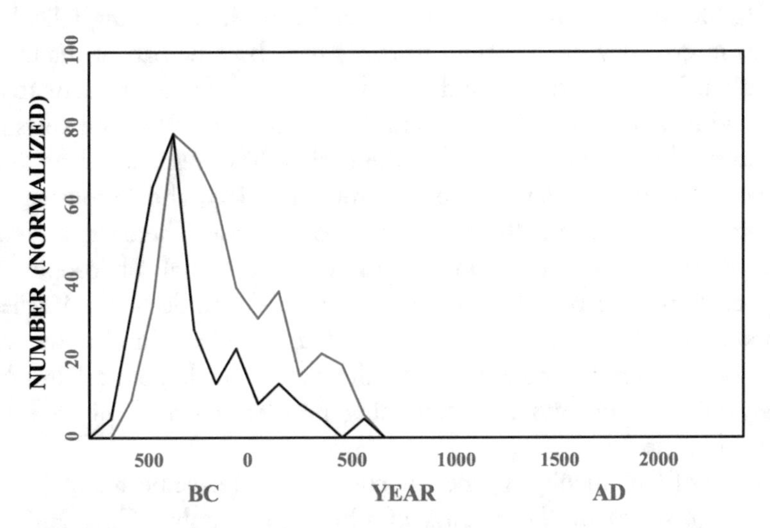

Fig. 2.1 Normalized numbers of Greek natural philosophers (black line) and other Greek philosophers (blue line) as a function of time. The total numbers are 61 (natural philosophers) and 397 (other philosophers). Refs: Greek natural philosophers (Bertman 2010), other Greek philosophers (Internet sources)

Although the Roman Empire overcame Greece in the second century BC, it did not put an absolute end to the Greek interest in abstract thinking, but neither did it adopt the Greek tradition itself. Some Romans took an interest, but overall there was mutual disdain—the Greeks looked down on the cultural inferiority of Rome, and most Romans saw Greek thinking as subversive and dangerous for the Empire.

Nevertheless, many Greek works were eventually translated into Latin. The pure science of the Greeks influenced the applied science of the Romans. Some Roman poets were inspired by the Greek classics. Other Romans taught liberal arts based on Greek writings, and still others wrote on scientific topics with reference to the Greek thinkers. There were a number of philosophical schools spread throughout the Roman Empire, but no more original thinkers. So overall the great intellectual culture initiated by the Greeks was not furthered by the Romans.

Why did the Greek tradition in natural philosophy finally come to an end? This question has long been discussed. Was it because it was thought that everything that could be done had already been done, and that the greatness (the 'received wisdom') of the philosophers of the past could never be surpassed? Perhaps it was the fact that much of what had been written was qualitative and speculative (with a few exceptions such as the *Almagest*) with no practical impact on human development—it could not actually be *used* for anything, so that, aside from pure interest, it was considered to be irrelevant to ongoing life. If it had been quantitative and capable of making reliable predictions in everyday life, it would certainly have become part of the fabric of civilization. World history would have been very different. But for that the world had to wait until the Scientific Revolution over a thousand years later.

Various other possible reasons for the decline of Greek natural philosophy have been proposed. It had no obvious role in society, and various cults and sects, which had no sympathy with it, were on the rise. There was an increase of scepticism and superstition, opposition to natural philosophy itself (as opposed to other branches of philosophy), a general decline in intellectual life, and a lack of continuity between teacher and student. In the later periods of the Roman Empire conditions rapidly deteriorated; political instability, military crises, epidemics and decreasing prosperity would all have had a negative impact.

The rise of Christianity is generally considered to have been a major factor in the later stages of the decline of Greek philosophy. Christianity first appeared as an obscure cult emerging out of the east in the first century, and the Christians were famously persecuted throughout the Roman Empire. But in a stunning reversal, in 380 Christianity became the official state religion of the Roman Empire. Christians had gained the upper hand and were then free to persecute the 'pagans'. A Christian imperial decree in the same year ordered that all pagan institutions in the empire be closed. That included the Great Library and Museum in Alexandria. According to some stories it, along with its hundreds of thousands of precious scrolls, was burned and destroyed by fanatical Christians, but there were also invasions by Syrians and Arabs. In any case it was gone by the early fifth century. The closing of the famous Platonic Academy in Athens by Imperial decree in 529 was another major blow. But it is obvious in Fig. 2.1 that the decline of Greek philosophy[1] was already underway centuries before the time of Christ, so Christianity could only have had an effect in the later stages of decline. Nevertheless, clearly the effect was devastating, and it marked the definitive end of the Greek golden period.

2.4 In Transit

In 285 it was realized that the Roman Empire had become too big for central control. It was subdivided, and the western part, later known as the Holy Roman Empire, remained under the administration of Rome, while the Eastern Roman Empire, later known as the Byzantine Empire, was to be governed from Byzantium (Constantinople). This distinction ultimately resulted in two very different histories, even though the two halves were reunited a few times. The Western Roman Empire, which included Western Europe, underwent various political and economic upheavals, and eventually disintegrated. Rome itself was sacked by a succession of invaders, including the Gauls in 387, the Visigoths in 410 and the Vandals in 455. The Western Roman Empire is traditionally deemed to have ended in 476.

At that point the Dark Ages had descended upon Europe. It became a scene of desolation, a cultural and economic backwater with scattered agriculture, tribes and towns (the Roman cities could not be sustained, and became depopulated). By the sixth century the only remnants in Europe of the Greek Golden Age were occasional fragments in isolated monasteries. Most of the inhabitants were illiterate. Europe seemed a very unlikely place to inherit the mantle of high civilization from Greece.

[1] Note that Figs. 2.1, 2.2 and 4.1 are only indicative and undoubtedly incomplete, but the major features are unlikely to be due to selection effects, in particular the declines following the peaks, as more recent scientists are more likely to be known to us than earlier scientists.

In contrast, elsewhere in the world there were great contemporary empires that vastly outshone Europe. These included Byzantium, greater India, China, Mesoamerica and South America. China was unparalleled. The Chinese empires and dynasties were the largest in the world, and they provided thousands of years of cultural continuity. In 1200 the population was 115 million people, with five cities over a million. Major achievements included not only the Great Wall, but also the Great Canal, at 1800 km the longest in the world. At one time China had the largest navy in the world, with hundreds of ships at its peak (the famous expeditions in colossal ships in the early 1400s reached the coast of Africa, but were suddenly stopped by a new emperor who wanted to concentrate on China's internal development and security). The bureaucracy for such a large centralized state was obviously huge. It supported scientific activities—mathematics, cartography, geography, seismology, meteorology, astronomy, astrology, alchemy, medicine—but all of course for practical purposes. The list of Chinese innovations is long, and includes gunpowder (a by-product of alchemy), the magnetic compass, paper, printed books with moveable type, the wheelbarrow, the umbrella, advanced metallurgy, suspension bridges, the fishing reel, porcelain china, lacquer, textiles including silk, the mechanical crossbow, and the world's first seismometer. With its long history of astronomy, it provided the longest unbroken run of astronomical records in the world.

Why, then, did China not develop natural philosophy as the Greeks did? Why did it not have a scientific revolution? Unlike Greece, China was a single huge empire, controlled by a strong central authority. As in the other early civilizations, the massive central administration would generally have been authoritative, conservative and all-pervasive. The first Chinese emperor burned books and buried scholars. The bureaucrats who ran the country were educated in Confucianism, which was solely concerned with human affairs and statecraft. The Buddhists regarded the natural world as an illusion, and the Taoists would have considered the very thought of any 'order' in nature that could be discerned by mere mortals as naïve and anathema to the fundamental principles of Taoism. So there were several possible reasons for the absence of natural philosophy. Another possibility is that the idea of studying the intrinsic properties and workings of the natural world just didn't occur to anyone over all those millennia. In any case, for all its greatness, China, like the other bureaucratic civilizations mentioned above, lacked natural philosophy in the Greek sense, and never had a scientific revolution.

One might have expected that the Greek philosophical tradition would have been carried on in Byzantium, the eastern part of the Roman Empire. In contrast to the Western Roman Empire it did not collapse; on the contrary it survived continuously, with various ups and downs, for over a thousand years, until 1453. Greece itself was a part of the Byzantine Empire, and, conveniently, Greek was the dominant language. A significant number of the Greek classics survived and a number of Byzantine scholars were familiar with them. Education was fairly widespread, and there were extended periods of relative calm which would have been conducive to scholarship. There were 'schools' of philosophy, and students were instructed in the traditional liberal arts in preparation for careers in the state bureaucracy.

However, Christianity had a significant effect on scholarship, as in all other areas of Byzantine life. The emphasis was on reconciling 'paganism' with the dogma of Christianity, so the spirit of unlimited free thinking that had been so important to the ancient Greek philosophers was gone. Over the centuries, almost all of the Byzantine activity related to the Greek classics was just in the form of teaching, debates, commentaries, annotations, preservation and transmission. There was very little original work—certainly nothing to compare with that of the ancient Greeks. And natural philosophy itself was largely ignored.

One notable exception was John Philoponus (c. 490–570), a Christian who was highly critical of Aristotle. He rejected Aristotle's dynamics in favour of a 'theory of impetus' (anticipating the concept of inertia), he opposed Aristotle's claim that heavy bodies fall faster than light bodies, and he made early arguments in favour of empiricism. He had a major influence on Galileo, who referred to him over a thousand years later in his own works. But overall, aside from Philoponus and a few less influential commentators, there was no great blossoming of new ideas in natural philosophy, in spite of the thousand years of existence of the Byzantine Empire.

2.5 Islamic Science

A major new development was the rise of Islam starting in the seventh century. Baghdad was established as the capital of the Islamic empire under the Abbasids in 762, and soon its population swelled to over a million. The only book in Arabic at that time was the Qur'an, and studying it produced a more general interest in scholarship. The Qur'an stressed the religious duty of all Muslims to seek knowledge and enlightenment. Furthermore, it was realized that the Islamic world was far behind other empires in many respects, and the Abbasids had an obsession with Persian culture. There was also great interest in astrology. And practical pursuits such as agricultural engineering, the calendar and accountancy called for specialized knowledge. For these and other reasons, a massive 'translation movement' took place over two centuries: the wisdom of the earlier civilizations of the Greeks, Persians and Indians was translated into Arabic. Once the movement got started it was unstoppable. The elites of the Abbasid society in Baghdad competed with each other for the most significant manuscripts, both for prestige and for the practical benefits they could bestow.

The Abbasid historical period lasted until the Mongol conquest of Baghdad in 1258, and is considered to have been the "Islamic Golden Age". Translations of nearly all the works of Greek natural philosophy known today were made at that time. Furthermore, following the Moorish conquest of Spain beginning in the eighth century, Muslim scholars brought these works with them and a golden age of Islamic civilization took place on Spanish soil.

As more and more translations became available, local original research inevitably arose—scientific activity became part of the Islamic world. Unlike the early civilizations and their polytheistic religions that claimed to explain everything, Islam did not present an impenetrable obstacle that made science impossible. On the contrary, the Qur'an encouraged Muslims to study nature, considered science to be important and emphasized the need for evidence and proof. There was a sacred and holistic aspect to these views, as nature, humanity and the cosmos were all seen as the works of God Himself.

Islamic science, including astronomy, mathematics and medicine, became a world leader for a period of several centuries, with activity ranging all the way from central Asia to Iberia. The 'sudden' appearance of so many of the Greek classics undoubtedly accounts for the steepness in the rise of Islamic science, which is clear in Fig. 2.2.

It is interesting to ponder why so much original work in natural philosophy was stimulated by the Greek classics in the Islamic world, but not in the Byzantine Empire, which existed at the same time. Perhaps it was because the Greek classics were totally new and exciting for the Muslim scholars, whereas the Byzantines may have regarded natural philosophy as part of a dreary standard curriculum no longer capable of generating any great excitement.

One of the first of the great scientists of Islam was the chemist Jabir ibn Hayyan (c. 721–815), who lived in Kufa, not far from Baghdad. His chemical research and scientific method were advanced for their time, and he became known in Europe as Geber the Alchemist. Al-Khwarizmi (Latin name Algorithmus) (c. 780–850), a mathematician, astronomer and geographer in Baghdad, was best known for writing the first book on algebra, *Kitab-al-Jebr*. He was influential in making the Hindu decimal system known in Europe as well as the Islamic world. Also in Baghdad, al-Kindi (Latin name Alkindus) (c. 800–873) became known as 'the Philosopher of the Arabs'. He worked and wrote on physics, astronomy, mathematics, medicine, music, pharmacy and

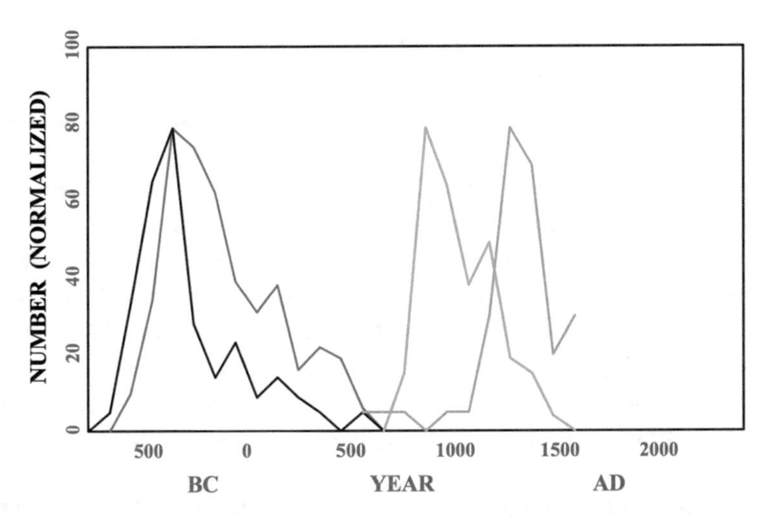

Fig. 2.2 Normalized numbers of philosophers and scientists as a function of time in the Greek, Islamic and medieval periods: Greek natural philosophers (black line), other Greek philosophers (blue line), Islamic scientists (green line), and medieval scientists (brown line). The total numbers are 61 (Greek natural philosophers), 397 (other Greek philosophers), 75 (Islamic scientists) and 51 (medieval scientists). Refs: Greek philosophers as in Fig. 2.1; Islamic philosophers (Al-Khalili 2012); medieval scientists (Freely 2012; Grant 1996)

geography, and was active in the translation movement. The Persian physician, philosopher and chemist Ibn Zakariyya al-Razi (Latin name Rhazes) (c. 854–925) was born in the ancient city of Rayy (part of present-day Tehran), and worked both there and in Baghdad. He ran several hospitals, was an early proponent of the scientific method, and wrote two medical books which became among the most important in medieval Europe. At the other end of the Islamic world Abu al-Qasim al-Zahrawi (Latin name Abulcasis) (c. 936–1013) lived near Cordoba, and was the greatest surgeon of Islam and the medieval world. He wrote a medical encyclopaedia which contained sections on surgery and surgical instruments; it was translated into Latin, and made him famous across Europe.

The three greatest scientists of Islam all lived and worked between 950 and 1050, at the peak of the Islamic Golden Period. Ibn al-Haytham (Latin name Alhazen) (c. 965–1039) was born in Basra and worked in Egypt. He was a physicist, probably the greatest over the entire period between the ancient Greeks and Galileo. His book on optics had a huge influence on Western science, and he made important contributions to astronomy. He was also an early proponent of the scientific method. The Persian al-Biruni (973–1048) was born in the city of Khiva and travelled and worked throughout Central

Asia. He was one of the great polymaths in history—philosopher, astronomer, mathematician, geographer and anthropologist. He wrote several important books, and became famous for measuring the size of the Earth to unprecedented accuracy. Another Persian scholar, Ibn Sina (Latin name Avicenna) (980–1037), was born near Bukhara, lived and worked in central and western Asia, and died in Hamadan. He is considered to have been by far the most influential and important thinker in Islam, and one of the most important in history—almost as influential in Western philosophy as Aristotle himself. He is best known as a philosopher and physician, and wrote important works in both fields. He also contributed to mathematics and physics, including concepts of light, heat, motion and infinity. It has been said that his contributions to science were so great that they actually discouraged further study in those fields.

Yet another notable Persian scholar was Omar Khayyam (1048–1131), one of the great mathematicians of the Medieval Period. He was born and worked in Nishapur. His contributions included work on cubic equations and calendrical studies. Also Persian, al-Ghazali (Latin name Algazel) (1058–1111), was born in Tus and worked in Nishapur and Baghdad. He was the most famous theologian in Islamic history. He made contributions to science, but was best known for his attacks on Aristotelian philosophy and its proponents such as Ibn Sina, and he was blamed (rightly or wrongly) for the decline of the golden age of Islamic science. Ibn Rushd (Latin name Averroes) (1126–1198), born in Cordoba, was one of the best-known philosophers of medieval times. He, more than anyone, was responsible for introducing Aristotelian philosophy to Europe. He was the last of the great Muslim philosophers, and deeply influenced Western thought. Nasr al-Din al-Tusi (1201–1274) was another Persian polymath who, like al-Ghazali, was born in Tus. He contributed to astronomy, physics, mathematics, chemistry, biology, medicine and philosophy. His life was disrupted by the Mongol invasion, but he later persuaded the Mongol leader to build him an observatory in Maragha, which became the most important astronomical centre for several centuries.

After peaking around 900–1000 AD, Islamic science began to decline. Its rise and fall are illustrated in Fig. 2.2. Conservative religious forces became less tolerant of 'foreign' studies, and the colleges known as madrassas restricted their curricula to religion, excluding philosophy and science. Individuals like al-Ghazali added to the discontent. In 1194 scientific and medical texts were burned in Cordoba by the Ulama. And the Arab world was under attack from the Mongols and Crusaders. In 1258 the Mongols destroyed Baghdad, and most of the books in it. Conditions for free and enlightened scholarship had evaporated. But by then many of the most important Greek classics had been passed on to Europe, where the torch would continue to burn.

2.6 Medieval Science

The first seeds for medieval science in Europe had already been put in place as the last embers of the Roman Empire were still smouldering. Translations of Greek classics into Latin were being made by Boethius (c. 480–525) and Cassiodorus (c. 490–580). Isadore of Seville (c. 560–636) wrote the first European encyclopaedia, which became popular throughout medieval Europe. The first Irish monasteries became important in spreading scholarship to others. The Venerable Bede (674–735) wrote manuscripts for monastic schools with some practical texts and some original scientific work as well as his views on history, cosmology, astronomy, natural science, mathematics and the calendar. His works were major sources of knowledge for centuries.

In 789 Charlemagne established monastic and cathedral schools throughout his domain, giving rise to a wide-ranging intellectual revival known as the Carolingian Renaissance, which helped to set the scene for European science. Gerbert d'Aurillac (c. 945–1003) was the first major intellectual figure in this tradition, which emphasized practical science as well as the classics. The translation effort from Greek and Arabic into Latin was accelerating, giving access to the Greek classics and original contributions from the Islamic world as well. Some of the important translators were Constantine the African (fl. 1065–1085), Adelard of Bath (fl. 1116–1142), Raymond of Toledo (fl. 1125–1151), and Gerard of Cremona (1114–1187). By the end of the twelfth century most of the available Greek and Arabic texts had been translated into Latin, and this body of knowledge formed a large part of the curricula at the new European universities. All these developments produced a revival that has been called the Twelfth-Century Renaissance, and European science was suddenly on the rise (see Fig. 2.2).

The first European universities were in Bologna (1088), Paris (1150), Oxford (1167), Cambridge (1209), and Padua (1222). By 1500 there were as many as 80 universities across Europe—a remarkable rise in the intellectual capacity of Europe. The university was an important and unique institution; none of the great civilizations had anything like it. It has been estimated that well over half a million students graduated from the European universities over the period up to 1500.

As a major part of the curricula at these universities was devoted to the works of Aristotle, there was inevitably a clash with the dogma of the Roman

Catholic Church. In 1210 the reading and teaching of Aristotle's works on natural philosophy at the University of Paris were banned for a period of 40 years on penalty of excommunication, and 219 specific works were condemned in 1277. Aristotle held that the universe was eternal and that events were determined by cause and effect, whereas Catholicism held that God had created the universe and that He could determine what happened at any time through divine intervention and miracles. The Church also objected to Aristotle's pantheism, the view that God is nature itself rather than a caring anthropomorphic God. And Aristotle's geocentric crystalline spheres got in the way of the biblical Ascension into the heavens. Two scholars struggled to reconcile these contrasting views: Albertus Magnus (c. 1200–1280) and Thomas Aquinas (c. 1225–1274). The views of both were strongly opposed by Catholic theologians. Thomas in particular struggled to find a compromise, and ended up by Christianizing Aristotelianism and Aristotelianizing Christianity. From that point on, 'Thomism' came to be the official position of the Catholic Church (and Thomas was made a saint in 1323). This meant that some Aristotelian doctrines such as the geocentric worldview became frozen in as Church dogma, and this posed a major problem that had to be overcome a few hundred years later in the Scientific Revolution.

The intellectual revival of Western Europe continued, and in some cases went beyond the level achieved by the Greek thinkers. Robert Grosseteste (c. 1175–1253) was an influential figure in England, developing the scientific method of verification and falsification, and exploring a wide range of topics including optics, acoustics, astronomy, calendar reform, mathematics, and (like many others) the rainbow. Roger Bacon (c. 1220–1292) was strongly influenced by Grosseteste at Oxford; he too had a wide range of interests, and wrote on the principles of experimental science. Peter Peregrinus (fl. 1265) studied the properties of magnets, and published his work in an influential treatise *De Magnete*. Another early European scientist was Jordanus Nemorarius (fl. 1120), best known for his work on statics (the study of forces in equilibrium). Grosseteste's scientific methodology was extended by William of Ockham (c. 1285–1349), who stressed the desired economy of scientific explanations in his famous 'Ockham's razor': "Entities must not be multiplied beyond necessity" (i.e. the simplest theory is preferred). This remains a guiding principle even today, for good reasons.

A number of thinkers studied dynamics, anticipating Newton's work a few hundred years later. These included the Oxford group of Thomas Bradwardine (c. 1290–1349), William Heytesbury (fl. 1330–1340), John of Dumbleton (fl. 1331–1349), and Richard Swineshead (fl. 1340). In Paris Jean

Buridan (c. 1295–1358) produced an 'impetus theory' which came close to Newton's laws of motion, and it became well known throughout Europe. Buridan's work was followed up by one of his students, Nicole Oresme (c. 1320–1382), who had prescient ideas in both dynamics and astronomy. The study of optics continued through the work of Witelo (c. 1230–1275), John Pecham (c. 1230–1292), and Dietrich of Freiburg (c. 1250–1311). Several influential texts on astronomy were produced that supplemented and augmented those from Greek and Arabic sources, including those by Grosseteste, Johannes de Sacrobosco (fl. 1220), and Campanus of Novara (fl. 1260). Other contributors to astronomy in the Medieval Period included William of St. Cloud (fl. 1290), John of Sicily (fl. 1290), John of Saxony (fl. 1330), Levi ben Gerson (1288–1344), and Richard of Wallingford (c. 1292–1336).

In 1347–1350 the plague known as the Black Death devastated Europe, killing a third of the population, and it brought many of these scientific activities to a sudden end. It recurred intermittently for centuries, and intellectual activity only gradually recovered. The Black Death probably accounts for much of the significant decline shown in Fig. 2.2. Some of the more well-known scholars during this period of recovery were Nicholas of Cusa (1401–1464), Georg Peurback (1423–1461), and Johannes Regiomontanus (1436–1476). Certainly the best known of all was Leonardo da Vinci (1452–1519). He was the ultimate 'Renaissance Man' who, in addition to his magnificent art, was interested in an extraordinary variety of subjects including astronomy, optics, mathematics, anatomy, botany, zoology, geology, astrology, alchemy, hydrodynamics and engineering. In many areas he was far ahead of his time.

Meanwhile, Johannes Gutenberg (1398–1468) introduced the printing press and moveable type to Europe in 1439—a monumental innovation that underpinned other major developments that all led to the modern world. It was one of the most important events in history. In 1492 Christopher Columbus (1451–1506) found the New World, opening up whole new vistas of discovery for Europe. And in 1517 Martin Luther (1483–1546) started the Protestant Reformation. All of these breathtaking developments dramatically changed Europe, broadened its horizons, and encouraged open and innovative thinking.

It had taken over a thousand years following the destruction of the Great Library of Alexandria for Europe to reach this point. The great works of the ancient Greek philosophers and the Islamic and medieval scholars were now all known, printed and well-studied, and the excitement stirred by the extraordinary developments of the Renaissance was inspirational. The scene was now set for a major turning point in the history of science.

2.7 The Scientific Revolution

The second of the two fundamental steps in the Rise of Science occurred in the sixteenth and seventeenth centuries in Europe—the Scientific Revolution.

Nicolaus Copernicus (1473–1543) was born in the Polish town of Torun. He attended the University of Krakow, and was exposed to the major intellectual works available, from antiquity to the modern works of the time. He then went to Renaissance Italy and studied law, medicine and the classics at the Universities of Bologna, Padua and Ferrara. In 1506 he returned to Poland, where he became a canon at Frombork Cathedral, practiced medicine, and carried out various civil duties for the rest of his life.

During his studies he was exposed to astronomy, and developed an interest in the ancient Ptolemaic model for computing the positions of the Sun, Moon and planets. He became aware of some of the major shortcomings of that model, in particular its difficulty with the 'retrograde motions' of some of the planets, its failure to explain why Mercury and Venus never venture far from the Sun, its incorrect prediction that the Moon should vary significantly in apparent size, and generally its cumbersome structure with epicycles, deferents and equants. He wondered whether it might be simpler and more elegant if one considered the Sun, rather than the Earth, to be at the centre of the universe.

He was not a revolutionary. He was looking to improve, not replace, the Ptolemaic system; for example, he maintained the circular motions that were basic to the classical model. But his heliocentric model immediately solved some of the problems. The peculiar motions of the outer planets were automatically explained by the different orbits of the various planets as they moved relative to each other going around the Sun. Mercury and Venus were closer to the Sun than the Earth, explaining why they never appear far from the Sun on the sky. The Moon was the one body that did orbit around the Earth, explaining why its apparent size does not vary considerably. So while it is true that the Copernican model still required epicycles and was not very much simpler (or more accurate) than the Ptolemaic model, it solved some of the problems and was certainly more elegant.

But, like the heliocentric model proposed by Aristarchus over 1500 years previously, it had a few problems of its own. The most serious was that the observed relative positions of the stars should change as the Earth orbits around the Sun (at opposite sides of the orbit the stars are seen from different angles). Both Aristarchus and Copernicus answered that this effect would be too small to detect if the stars are at extremely large distances from the Earth. And there were the 'common-sense' issues: why do things not fly off the spinning Earth as it moves around the Sun, why do we feel no great constant wind from the east, why do objects fall straight down onto the Earth rather than towards the west, why do birds have no more difficulty flying east than west, why does a cannonball go the same distance east or west? The answer given to some of these questions was that objects on or near the surface of the Earth share in its motion around the Sun.

So Copernicus was not in a rush to publish his theory. In addition to the possible criticisms it would face, it could also prove highly unpopular with both the Roman Catholic Church and the Protestant Church, for which the Earth is forever at the centre of the universe as a matter of religious dogma. Copernicus tentatively published an anonymous tract in 1514, which he distributed to some of his closest friends. Towards the end of his life he was finally convinced to publish a book, *De revolutionibus orbium coelestium (On the Revolution of the Heavenly Spheres)*, dedicating it to the Pope. The book was published in 1543, the year he died.

For such a 'revolutionary' book,[2] it received little attention for many decades. One reason was that it was rather technical. It was read by various astronomers, and one of them even computed a new set of astronomical tables based on it. But it was not until the seventeenth century that his work became famous and the Roman Catholic Church banned the book and any teaching of the works of Copernicus—that story will unfold shortly.

A few decades after the time of Copernicus, astronomical measurements of much greater accuracy became available. Danish astronomer Tycho Brahe (1546–1601) built a large and comprehensive observatory north of Copenhagen, and conducted a series of observations of unprecedented accuracy over a period of two decades. In 1600 he moved to Prague under a new patron, and hired an assistant by the name of Johannes Kepler (1571–1630). Kepler was a young German mathematician whose work in mathematical astronomy had attracted the attention of Tycho. Unlike Tycho he was a supporter of the heliocentric model. Soon after Kepler arrived Tycho died, and Kepler was appointed head of the observatory.

Working with Tycho's extensive data over several years in the context of the heliocentric model, Kepler discovered his three laws of planetary motion: the

[2] The modern meaning of the world 'revolutionary' derives from the title of this book.

planets move in elliptical orbits around the Sun (not circles), the line from the Sun to a planet sweeps out equal areas in equal times, and the square of the period of a planet's orbital motion is proportional to the cube of its average distance from the Sun. These three laws provided the astronomical basis for Newton's theory of universal gravitation which was to come later in the century. In 1627 Kepler published a new set of tables, the *Rudolphine Tables*, using Tycho's accurate observations, a heliocentric model of the solar system, and his knowledge of the elliptical orbits of the planets. The agreement of the model with the observational data was remarkable; this was a huge step forward in astronomy.

* * *

Galileo Galilei (1564–1642) was one of the key figures in the Scientific Revolution, and one of the greatest scientists in history. He was the first modern scientist, using mathematics to describe the motions of objects and using experiments to test hypotheses.

His early years were mixed—he joined a monastery at 15, became a medical student at the University of Pisa at 17, then fell in love with mathematics and dropped out of university to become a tutor of mathematics and natural philosophy. He had an inquisitive mind, he loved to argue, and he questioned the Aristotelian wisdom of the time. Unlike Aristotle he thought that all objects fall at the same speed. At an early age he discovered that the period of a pendulum's swing depends only on its length, not on its weight or the length of its arc.

He returned to the University of Pisa as a professor of mathematics in 1589, and his fame as an independent and intelligent thinker gradually spread. In 1592 he was appointed to the chair of mathematics at the University of Padua, a part of the Venetian Republic. He recalled the 18 years he spent there as being the happiest in his life.

He worked on various practical problems, including military fortifications, how systems of pulleys work, and a rudimentary all-purpose calculator that he invented. He did his famous experiments with balls rolling down inclined planes, establishing that balls of different weight accelerate at the same rate due to gravity; he knew that the same would apply if the balls were in free-fall—this is just the extreme case in which the inclined plane is vertical. He had a "water clock" as a timer, measuring the accumulated water coming from a vessel with a hole in the bottom. His experiments were quantitative, in contrast to the qualitative science of the Greeks.[3]

Galileo thought about inertia, and imagined a ship at sea moving at constant speed; everything on the ship (balls being thrown, objects falling) behaves as if the ship is at rest. He considered what it is that keeps cannonballs flying through the air. In contrast to Aristotle he suggested that all bodies in motion tend to maintain that motion, and bodies at rest stay at rest; this became the law of inertia, and Newton's first law of motion. Galileo investigated hydrostatics and magnetism. He was a supporter of the Copernican model, and corresponded with Kepler and other scientists. He showed that a 'new star' that suddenly and astonishingly appeared as the brightest star in the sky in 1604 had no observable motion relative to other stars, and argued that it must be as far away as the other stars, challenging Aristotle's notion of an unchanging celestial sphere. (The 'new star' was in fact a much fainter star that had just ended its life in a spectacular explosion; such events are called 'supernovae'.) And he proved that the trajectory of a projectile fired from a cannon is a gently curving parabola, in contrast to Aristotle's view in which the projectile first travels horizontally and then drops to the ground.

It was in 1609 that Galileo first heard of the invention of the telescope by the Dutch spectacle maker Hans Lippershey (1570–1619). Copies of this early version, with a magnifying power of three times, were already being used as toys in fairs around Europe. Galileo immediately realized the potential military and trade importance of such an instrument to Venice, which made it possible to more easily see and identify distant ships. He quickly built a much more powerful telescope of his own, eventually reaching a magnifying power of 20–30 times.

He had the inspiration of pointing his telescope to the skies, and made astonishing discoveries: the brightest four moons of Jupiter moving in their orbits around Jupiter itself (proving that not everything just moves around the Earth), mountains and craters on the Moon (which is therefore not the perfect sphere that the ancients had imagined), and countless stars comprising the Milky Way. He published these in a short book *Siderius Nuncius* (*The Starry Messenger*) in 1610, which immediately became famous throughout Europe.

Shortly thereafter Galileo was offered a well-paying lifetime post as Philosopher and Mathematician to the Grand Duke of Tuscany. When he arrived in Florence he heard that Kepler had confirmed his discovery of the

[3] The ancient Greek philosophers wouldn't have stooped to anything as banal as repeatedly rolling balls down an inclined plane, but Galileo's experiments provided crucial information for Newton's law of universal gravitation.

four moons of Jupiter. Galileo subsequently also found that Saturn is not a perfect sphere, he discovered the phases of Venus (which could only be explained if Venus orbits the Sun), and he observed spots on the Sun, blemishes that revealed that the Sun is also not the perfect heavenly body it had been presumed to be.

These discoveries provided very strong direct evidence in favour of the Copernican model and against the ancient Ptolemaic model, but Galileo was nevertheless careful not to promote the Copernican model itself. He was well aware of the fate of Giordano Bruno (1548–1600), who was burned to death at the stake by the Roman Catholic Inquisition, in part for being a proponent of the heliocentric model. He maintained good relations with members of the Church hierarchy, including the Pope himself. There were still Aristotelian sceptics of Galileo's observations, suggesting that they may be due to artefacts of his telescope, but Galileo tested his telescope on nearby and known objects to show that this was not the case, and a Jesuit committee of the Church supported all the major observational claims made by Galileo.

But over the years Galileo started to express support for the Copernican model, despite his initial reservations. In 1616 a papal commission concluded that the idea that the Sun lies at the centre of the universe and that the Earth moves through space was heretical, and Copernicus's book *De Revolutionibus* was banned by the Church. The next year the Pope, and subsequently the Inquisition, decreed that Galileo must not support or teach the Copernican worldview. Galileo accordingly turned his attention to his interests on other scientific topics.

In 1623 a new Pope (Urban VIII) was elected who was much more positive about Galileo, and in the following year Galileo had six audiences with him and was awarded various honours. Most importantly, the new Pope granted him permission to write a book about the two systems of the world, the Ptolemaic and the Copernican, as long as he kept a neutral position. Galileo published the book in 1631, with the title *Dialogo sopra i due massimi sistemi del mondo (Dialogue on the Two Chief World Systems).* It was an imaginary debate between two advocates, Salvari (for the Copernican case) and Simplicio (for the Ptolemaic case). Unfortunately, the use of the name Simplicio implied that only a simpleton would believe the Ptolemaic worldview, and the text seemed to be sharply partisan in favour of the Copernican view. The Pope and the Church felt insulted. A papal commission was ordered to investigate, with the result that Galileo was ordered to stand trial before the Inquisition for heresy.

Again very aware of Bruno's fate, Galileo was finally forced to renounce his belief in Copernicanism ("I abjure, curse and detest my errors"). The initial sentence was life imprisonment, but it was later reduced to lifetime house arrest. Given that his house was a very fine villa in Arcetri on the outskirts of Florence, and that he could continue on with his other studies, it turned out not to have been such a bad outcome.

There he completed his masterpiece on physics, *Discourses and Mathematical Demonstrations Concerning Two New Sciences*. It covered his life's work on mechanics, inertia, the strength of materials, the nature of fluids, the weight of air, the behaviour of light, the flight of projectiles, the swinging pendulum, the loaded beam, and much more. It gave Galileo's law of falling bodies—the distance a body covers in free fall is proportional to the square of the time of the fall. And it discussed the scientific method: the importance of observations and repeated experiments to test hypotheses. It could not be published in Italy due to the ban on his writings, so it was smuggled out and published in Leiden in 1638. It had a huge impact on the subsequent development of science in Europe.

The treatment of Galileo by the Church had a severe dampening effect on the quantity and quality of science in Italy, which became a relative backwater for decades. But Europe, like Greece, had the advantage of being comprised of a variety of different regions and states. Galileo's findings and ideas were free to flourish in the rest of Europe, especially in the now-Protestant regions of the north which were no longer under the control of the Catholic Church. The Reformation was vital for the survival of science; if it had not taken place, Galileo's book would probably not have been published at all, and Kepler's and Newton's works would probably have been banned.

The Church finally lifted its ban on Copernicus' book *De revolutionibus* in 1758 and on Galileo's *Dialogue* in 1822, and in 1992, 350 years after Galileo's death, the Church finally pardoned Galileo.

* * *

Isaac Newton (1642–1726) was born into a farming family in Lincolnshire, England. He was raised, first by his parents, then by his maternal grandparents, and then at a grammar school. He is said to have been a lonely child. When he was 16 he was taken from school to manage the family farm. The experience didn't last long, as he had no interest in farming. At the age of 18 he was fortunate enough to be accepted into Trinity College, Cambridge. He eventually found that he could ignore much of the formal curriculum and spend most of his time studying whatever he wanted, which included the great works of science and mathematics. In spite of his obsession with his own work, he managed to win a scholarship that would enable him to stay at Cambridge until 1668. This period was disrupted in 1665 by a plague that was raging at the time. The university was temporarily closed, and Newton returned to Lincolnshire for two periods until 1667, when the plague had finally run its course.

By his own account the years 1663–1668 were some of his most productive. Some historians have called the year 1666 Newton's *annus mirabilis*. The famous story of Newton being inspired to think of universal gravitation by seeing an apple fall from a tree refers to his time in Lincolnshire. Certainly Newton did have many of his great ideas in that period, but his work matured over a considerably longer period. His remarkable abilities and achievements at that time were rewarded with his being made Lucasian professor of mathematics in 1669, a position which he held until his departure from Cambridge in 1687.

Newton had a variety of scientific interests—optics, mathematics, mechanics and gravity—and he also worked for considerable periods on alchemy and interpretations of the Bible.

In optics he used a prism to split white light into the colours of a rainbow. He then reversed the process, making white light again, showing that white light is in fact a mixture of all the colours. He also showed, again using a prism, that the component colours could not be further reduced or changed, indicating that they are truly properties of light rather than artefacts of refraction. This was a brilliant experimental result, revealing one of the fundamental properties of light.

As a technical spinoff from his work on optics, Newton designed and built a reflecting telescope in order to avoid the complications resulting from light refracted through lenses. When this was shown to the Royal Society in London in 1671 he was immediately elected a Fellow of the Society, and his famous paper on optics was published in 1672. He thought of light as a stream of particles (which he called corpuscles), in contrast to the views of others according to whom light is a wave phenomenon. One of these others was another member of the Royal Society, Robert Hooke (1635–1703), with whom Newton had various disagreements over the years. It was in a 1675 letter to Hooke that Newton famously declared "If I have seen further, it is by standing on the shoulders of giants".

Disputes and arguments over the details of his 1672 paper made Newton frustrated, and he became a recluse at Cambridge. For years his work developed, but he kept it largely to himself. He was obsessive, and worked long hours on his interests.

In 1684, Edmund Halley (1656–1742), Hooke, and Christopher Wren (1632–1723) were discussing the orbits of the planets at the Royal Society.

They suspected that Kepler's three laws might all follow if the Sun pulled the planets with a force that became weaker with distance in accordance with an 'inverse-square' law, but they couldn't prove it. Later that year Halley was in Cambridge, and visited the reclusive Newton. He mentioned the orbital hypothesis, and asked Newton whether he could prove it. Newton said he could, and by the end of the year he sent Halley the proof in a short paper entitled *De motu corporum in gyrum (On the Motion of Bodies in Orbit).* It was a stunning achievement, but Newton declined to publish the paper, saying he would rather study the matter further before publication. Over the next 18 months Newton made a Herculean effort that culminated in what is probably the greatest publication in the history of science.

Newton's *Philosophiae Naturalis Principia Mathematica* (Mathematical Principles of Natural Philosophy), published in 1687, was a monumental *tour de force* that changed science forever. Once and for all, the shackles of the ancient Aristotelian qualitative worldview were broken, to be replaced by a precise, quantitative and workable worldview of great power. The universe and all its contents work according to fixed laws[4] that make accurate predictions possible. It is a 'clockwork' universe. The implication was that if everything was known at one time, it would in principle be possible to predict everything into the future.[5] The change was absolutely revolutionary.

The *Principia* brought together all of the other great achievements of the Scientific Revolution—the heliocentric theory of Copernicus, Kepler's laws of planetary motion and Galileo's physical ideas, experiments and astronomical observations—into one monumental synthesis.

One of the crowning achievements of the *Principia* was to show that the law of gravity is universal—it applies both to distant planets and to objects on the Earth. In order to describe the orbits of the planets, Newton had to invent

[4] The modern concept of 'laws of nature' arose primarily in the seventeenth century. In ancient Greece 'laws' almost always referred to human conduct. The regularity of the movements of the heavenly bodies obviously inspired the Greek word 'astronomos' (whence 'astronomy'), combining 'astron' (meaning 'star') and 'nomos' (meaning 'law'), but that was a special case; Aristotle wrote about causes, not laws. There were various hints of the idea of laws of nature in medieval discussions involving providential theology, but it was René Descartes (1596–1650) who explicitly introduced the modern concept of laws of nature as we know it. It was truly innovative, and he struggled with it. In the end, he felt compelled to invoke God as underwriting these laws, both to provide forceful causal efficacy and to deflect any religious opposition (knowing what happened to Galileo), but soon the concept was able to stand on its own. It had a huge impact. When Kepler and Galileo produced their results (before Descartes' work) the word 'law' was not yet in use. But Newton came after Descartes, and he made full use of it: in the opening pages of his Principia are his three famous "Axioms, or Laws of Motion". In this he was certainly standing on the shoulders of Descartes.

[5] The first articulation of this view, called scientific determinism, was actually published much later, in 1814, by Pierre-Simon Laplace (1749–1827).

calculus, another major achievement. Using calculus, the orbits can be described as a series of infinitesimal segments that together form a smooth curve. The orbital motion of a planet is caused by its tangential motion being constantly deflected inward by the gravitational attraction of the Sun. In this way Newton was able to produce Kepler's three laws. And Newton's laws could make predictions of unlimited accuracy, both for planetary motions and for objects falling to the ground.

The *Principia* contains a general theory of force and motion. Newton's famous three axioms or laws of motion, expressed in modern terms, are simply (1) a body remains in a state of rest or constant motion if there is no net force acting on it, (2) the acceleration of a body is equal to the net force acting on it divided by its mass ($F = ma$), and (3) for every action there is an equal and opposite reaction.

In the *Principia* Newton showed that a wide variety of phenomena could be mathematically explained and predicted. These included the motions of two bodies around a common centre of gravity, the motions of the Moon, planets, comets, the moons of the outer planets and the Sun itself, why the same side of the Moon always faces the Earth, the origin of the tides of the sea, the figure of the Earth, the precession of the equinoxes, the motion of bodies in a resisting medium, the density and compression of fluids, hydrostatics, the motions of suspended bodies, the speed of sound in air, and the motion of fluids. From the motions of the planets to hydrostatics, all worked out with mathematical precision: there was no doubt as to how comprehensive this work was.

The *Principia* was instantly acknowledged as a book that would change the world. It was widely read, in Britain and all over Europe. Newton became famous. He dipped into politics and became a parliamentarian for a year. In 1700 he became master of the mint in London. His high salary enabled him to live in style, and he enjoyed a far more social life. In 1703 he became President of the Royal Society. He died in 1726, at the age of 84.

With the publication of the *Principia* the Scientific Revolution was complete. The works of Aristotle were relegated to the history shelves. The world was ruled by the immutable laws of physics.

This was the second of the two monumental steps in the rise of science. The *qualitative and descriptive* worldview of Aristotle was replaced by the *quantitative and predictive* worldview of Newton, which *worked*—and with astonishing accuracy.

The combination of Newton's laws, with their ability to make precise mathematical predictions, together with the scientific method—the use of experiments and observations to rigorously test such predictions—provided an extremely powerful way of understanding and predicting the world. It was truly revolutionary. It was a new way of thinking about the world, and a huge step for civilization. Modern science was born.

Science had received a huge boost, and was on the move on an increasing number of fronts. The development of science following the Scientific Revolution became exponential, and it would require a complete library to describe it all.

In the following brief overview, a few of the most important areas of science since the Scientific Revolution are followed to give an idea of the Rise of Science generally over the last few hundred years. These are The Very Large (out to the universe), The Very Small (in to the atom), Light (from optics to photons) and Life Itself (from botany and zoology to DNA).[6]

2.8 The Very Large

What is the biggest entity? While Newton's *Principia* finally settled the issue of the motions of the planets, there were many questions that remained. How big is the solar system? What are the comets that appear from time to time? What are the stars? How far away are they? Are they concentrated in a shell just outside the solar system, or do they extend out to infinity?

Using Kepler's laws, astronomers knew the *relative* distances of all the planets from the Sun, based on their orbital periods. But the *absolute* scale of the solar system was harder to determine. Giovanni Cassini (1625–1712) used the method of 'parallax' to determine the distance of Mars from the Earth. In 1672 he sent his colleague to French Guiana, while he stayed in Paris. They made simultaneous observations of Mars, and used the angle between the two lines of sight to determine the distance to Mars. As the relative distances of all the planets from the Sun were known, it was only necessary to determine one of them to know them all. The Astronomical Unit (AU), the distance from the Earth to the Sun, is the standard distance scale used in describing the solar

[6] It should be noted that these short histories are written with the benefit of '20–20 hindsight', looking back into the past and including just some of the scientific work that we now know led to our present knowledge in these four areas. They are therefore selective, and over-simplify the actual history of science to some extent. At any time in the past there were different avenues to follow, and it would not have been clear at that time what they would lead to, including detours and dead-ends. Science is as complex as any human endeavour, as stressed in the next chapter. This is also made clear in the many excellent and comprehensive books in the vast literature on the history of science. Finally, it should be noted that, while a large fraction of the developments after 1900 were honoured with Nobel Prizes, these awards are not mentioned here as they would clutter and interrupt the story lines; prizes such as these are discussed in Chap. 4.

system. Cassini's calculations gave 138 million km for the AU, which is fairly close to the currently accepted value of 150 million km. The solar system is big.

In 1664 James Gregory (1638–1675) pointed out that observations of the transit of Venus across the face of the Sun, made from different points on the Earth, could also be used to determine the distance from the Earth to the Sun using simple geometry. Edmund Halley followed this up later, predicting that the next transits would occur in 1761 and 1769, and giving details on how the observations should be made. As these two transits only occur every 243 years, this was an opportunity not to be missed. Not only would they provide the scale of the solar system, they would also give information relevant to the problem of determining longitude at sea. A further enticement for nations to take part in this enterprise was national prestige.

It was the first truly international scientific collaboration, and it was huge. To make it possible, the scientists from several different countries had to communicate, collaborate and agree on a plan. As some of the countries involved (such as Britain and France) were at war at the time, they granted safe passage to the nationals of their rivals. In 1761, some 120 observers from nine countries went to 62 posts around the world. Many of them had to endure severe hardships for months in getting to their appointed positions. Unfortunately this first transit effort was largely unsuccessful due to poor weather, so the stakes were very high for the 1769 transit. For this second transit Captain James Cook (1728–1779) took the HMS *Endeavour* on his famous trip to Tahiti to make the observations there, and the French government instructed its navy not to interfere with it, as it was "on a mission for all mankind". This time there were some 250 observers from ten countries at 130 locations, which included Europe, Siberia, North America, India, St. Helena, South Africa, Indonesia, China, as well as Tahiti. In spite of some initial disappointments the final results were impressive: they gave a value of 150.8 million km for the AU, within 0.8% of the correct value of 149.6 million km.

In the midst of all this, there were early indications that the speed of light is finite.[7] Observations showed that eclipses of the moons of Jupiter happen 'ahead of schedule' when Jupiter is closest to the Earth, and behind schedule when Jupiter is away on the far side of the Sun. The time taken for the light to reach the Earth depends on how far away Jupiter is—light takes time to travel. This insight was due to the Danish astronomer Ole Rømer (1644–1710), who

[7] The speed of light is 300,000 km/s. It is a universal constant, and nothing travels faster than light *in vacuo*. Astronomical distances are often expressed as multiples of a 'light-year', the distance that light travels in one year; that distance is 9 trillion km.

was working at the Paris Observatory; his estimate of the speed of light was within 25% of the true value, but his conclusions were controversial at the time. Another indication of the finite speed of light came from the discovery by James Bradley (1693–1762) of the 'aberration of light': the observed direction of the incoming light received on Earth from a distant star is changed slightly by the motion of the Earth as it orbits the Sun; the effect depends on the speed of the Earth relative to the speed of light. This also gave direct proof of the Earth's motion around the Sun.

Newton had suggested that the mysterious comets may just be bodies that move in highly elongated ellipses about the Sun, in which case they may reappear from time to time. With this in mind, Halley combed the historical records for appearances of comets with similar orbits separated by similar periods of time, and found one intriguing case in the comets of 1531, 1607 and 1682. He thought that these could be reappearances of the same comet (and that the difference between the intervals could just be due to perturbations by planets along the path), in which case the next appearance would be in late 1758 or early 1759. Subsequent detailed calculations taking the perturbation by Jupiter into account predicted that the comet would come closest to the Sun within a few weeks of mid-April 1759, and indeed this was observed on 13 March 1759—the return of 'Halley's comet' provided a stunning confirmation of Newton's physics.

Navigators at sea could easily determine their latitude by measuring the elevation of the Sun at noon, but determining longitude was a major problem. Until the late eighteenth century there was no solution, causing much hardship and the loss of countless ships and lives—they were traveling 'blind' west or east. Determining longitude required comparing the local time with a 'standard time' at the home port. How was it possible to know the time at that distant place? In the eighteenth century there were two competing possibilities—using the Moon's motion against a background of fixed stars as a celestial clock, or 'carrying' the standard time on board in a reliable chronometer (a precision clock) that is accurate even on rolling seas. As this was of major importance for the British navy and merchant fleet, a large prize was offered by the British government for the best solution. Many other European maritime nations were also involved in trying to solve the problem as it affected them all, and the effort became one of the largest scientific endeavours ever undertaken. Amazingly, it was the chronometer that won. Clockmaker John Harrison (1693–1776) made a succession of astonishingly sophisticated and compact chronometers, accurate to 5 s over 80 days, and, after a transatlantic trial of his best chronometer in 1764 he ultimately won the prize.

Much time and effort was devoted to making improved catalogues of the stars, giving their positions and brightnesses to provide navigators at sea with an accurate star catalogue. The Royal Greenwich Observatory was established for this purpose in 1675, and the first result was a catalogue of 3000 stars published in 1725. With a view to facilitating the determination of longitude using the Moon, the first of the annual Nautical Almanacs appeared in 1766. Similar work was underway in other countries. When the chronometers were selected as the preferred method to determine longitude at sea, the observatories established at the major ports around the world provided navigators with accurate astronomical calibration of their chronometers before they set sail.

Several other catalogues were prepared for purely scientific purposes. William Herschel (1738–1822) made a 2-m long 'backyard' telescope to look for 'double stars' (stars that are close to each other on the sky), and published catalogues of hundreds of them in 1782–1784. He was able to confirm that they are binary stars orbiting under mutual gravitational attraction. He later built a 6-m long reflecting telescope to survey the sky looking for nebulae (diffuse regions of light) and clusters of stars, and ended up with catalogues classifying thousands of them. His work was continued by his son John (1792–1871), and the combined effort culminated in the *New General Catalogue* containing 7840 deep sky objects, published in 1888.

The nature of the nebulae was uncertain—are they all distant star clusters masquerading as single diffuse objects? William Parsons (1800–1867), the 3rd Earl of Rosse, built a giant telescope with a 1.8-m diameter mirror to answer this question in 1845. It revealed that some nebulae have a spiral structure, and it resolved stars in the Orion Nebula, but in the end it did not give a definitive answer as to the nature of the nebulae—that had to wait for spectroscopy.[8]

Spectroscopy first became a tool in the 1800s. In 1802, William Wollaston (1766–1828) repeated Newton's famous experiment with sunlight, but instead of a pinhole he used a narrow slit 1.3 mm wide, and to his surprise he found seven dark lines in the spectrum. The German optician Joseph Fraunhofer (1787–1826) invented the spectroscope in 1814, and with it he discovered a bright orange line in the spectrum of a fire, and 576 dark lines in the solar

[8] In spectroscopy we spread out the spectrum of light into its component colours (wavelengths), from blue to red, using a prism or spectroscope. Different sources of light can produce narrow, bright (emission) or dark (absorption) features at specific wavelengths in the spectrum. These are due to atoms and molecules in the source of the light, and are referred to as emission and absorption 'lines' (a 'line' is just an image of the spectroscope's slit). Their relative strengths tell us the chemical composition of the source. The lines can also be shifted along the spectrum by the motion of the source. If the source is moving towards us, the lines are shifted towards the blue end of the spectrum, and if the source is moving away from us the lines are shifted towards the red. These are called blueshifts and redshifts, and this well-known phenomenon is referred to as the 'Doppler effect'.

spectrum. By 1859, two other Germans, Wilhelm Bunsen (1811–1899) and Gustav Kirchhoff (1824–1887) had established that hot solids and liquids produce a continuous spectrum (free of lines), while glowing gases produce a bright-line spectrum. In 1864, the astronomer William Huggins (1824–1910) found that one of the nebulae has a bright-line spectrum, proving that 'true' gaseous nebulae do exist. It was later found that when a continuous spectrum was passed through a gas, dark lines characteristic of that gas appeared. In astronomy these lines, bright and dark, are emission and absorption lines due to atoms in the stars and nebulae—the same atoms we observe here on Earth. Therefore, we can use our knowledge of the atoms we observe in the laboratory to determine what the distant stars and nebulae are made of!

The reverse is also true. During a solar eclipse in 1868, an unfamiliar pattern of lines was seen in the spectrum of the Sun by Pierre Jansen (1824–1907) and Norman Lockyer (1836–1920). Lockyer thought that they must be due to some as-yet unknown element, which he named 'helium' (from the Greek *Helios*: the Sun). Helium was subsequently identified on Earth 27 years later, in 1895. So it was proven that, just as Newton's laws of gravity apply both on the Earth and in the cosmos, the same is true for atomic physics. The matter in the distant universe is the same as the matter here on Earth! A stunning revelation.

Two new planets were added to the original six in the eighteenth and nineteenth centuries, one a chance discovery, and the other a prediction based on Newtonian mechanics. The discovery was made in 1781 by William Herschel in the course of a series of observations using his large telescope. He found an unusual object, which turned out to be the planet Uranus. It is likely that Uranus had been observed several times before (even by Hipparchus in 128 BC), but had not been recognized as a planet. Further observations indicated that its motion deviates from a simple planetary orbit. Possible reasons were considered, and it was concluded that gravitational perturbations by another massive object were most likely. Calculations in both England and France predicted the location of that object, and observations in 1846 found it and revealed that it was indeed another planet—Neptune. Another stunning success for Newton's physics.

What about the stars? Until the Scientific Revolution astronomers saw the stars as part of the outer celestial sphere, unchanging both in position and brightness. However, in 1572 Tycho discovered a supernova suddenly appearing brightly in the heavens and then gradually fading, and it was realized that stars might not be so uninteresting after all. In his influential seventeenth century cosmology, René Descartes (1596–1650) proposed that the stars are luminous objects like our Sun, only much more distant and scattered throughout the universe.

This new concept opened up a whole new field of study. Do the stars move? Do they vary in brightness? How are they distributed in space? To address the first question, Halley worked with an early version of the Royal Greenwich Observatory's catalogue of stars, and compared the stellar positions with those from a smaller catalogue assembled by Hipparchus in the second century BC. In 1718 He found that in a few cases the differences in the positions were huge—larger than the angular size of the Moon on the sky. This was far greater than could be explained by mistakes in the ancient catalogue, indicating that the stars had actually moved—further conclusive proof against a celestial sphere of fixed stars.

The observed rotation of sunspots suggested the possibility that other stars may also rotate, and as a result could vary in apparent brightness. But, aside from novae (stars that brighten suddenly and briefly, of which the 'supernova' discovered by Tycho was an extreme example), the expected variations would be small and hard to detect with the telescopes of the time. Nevertheless, some stars were found in the eighteenth century to have significant variations, in one case due to an eclipse, and in another case due to pulsations—one of the so-called 'Cepheids' that were to play a crucial role in twentieth century cosmology. But overall, a proper scientific study of such brightness variations had to await later developments.

How distant are the stars? In the heliocentric model the Earth orbits the Sun every year, so every half-year it is on opposite sides of its orbit, and observations of a given star will show it at different angles—it will appear to have moved (the so-called 'annual parallax'). The fact that this effect was too small to be observed meant that the stars had to be at extremely large distances, as both Aristarchus and Copernicus had argued. By the early 1700s, improved attempts to measure parallax put the distance to the stars at more than 0.4 million AU. A different approach to estimating the distances of the stars was to assume that the Sun is a typical star, and compare its brightness with that of the stars. As the apparent brightness of an object is inversely proportional to the square of its distance from us, the distances to the stars could be determined relative to the distance from the Earth to the Sun. In this way Christiaan Huygens (1629–1695) estimated that the star Sirius (the brightest in the sky) is 28,000 AU away from us. Newton used a variant of this method, putting Sirius at one million AU from us. He wasn't far off—we now know that Sirius is 0.54 million AU away.

But the holy grail in determining the distances of the stars was still to make *direct* measurements of the stellar parallaxes. This required the precise measurement of changes in the apparent position of a 'nearby' star relative to the background of distant stars due to the Earth's orbital motion around the Sun.

The effects are very small (fractions of a 'arcsecond', which is about one two-thousandth of the diameter of the Moon), and there were several complications to be dealt with. But finally, after more than a hundred years of effort by several astronomers, in 1838 Friedrich Wilhelm Bessel (1784–1846) reported a convincing measurement of a star with a parallax of 0.3 seconds of arc, implying a distance of 10 light-years. This is just one of the nearest stars, and it takes 10 years for its light to reach us! Finally, astronomers were becoming aware of the awesome size of the universe.

These early measurements had to be made by eye, and it was a slow business. By 1900 only about 60 parallaxes were known, but that was just enough to obtain useful statistics about the properties of stars. With the distance and apparent (observed) brightness of a star, it was possible to determine the true (intrinsic) luminosity. And using photographic plates to record the images and spectra of the stars, it became possible to determine their colour and composition.

One further important parameter was the mass of a star. This information was obtained from studies of binary stars—pairs of stars that orbit around each other. Spectroscopic observations give the velocities of the two stars using the Doppler effect. The masses of the stars in the binary system can then be computed using Newtonian physics. With these ingredients it became possible to study the physics of the stars—astrophysics was born.

A crucial relationship between two stellar parameters was found independently in 1910 by two astronomers, the Dane Ejnar Hertzsprung (1873–1967) and the American Henry Norris Russell (1877–1957). The parameters are the luminosity of a star and its colour (indicating its temperature). The graph showing this relationship is called an H-R diagram, and it remains one of the most important tools in the study of stars. Most stars lie along a so-called 'main sequence', with hot, massive stars at one end and cool, low-mass stars at the other. They depart the main sequence in characteristic ways as they become old, so the entire lives of whole classes of stars can be followed on the H-R diagram. The Sun itself is an average star, and it lies in the middle of the main sequence. The H-R diagram gave astrophysicists a powerful tool with which to understand the evolution of stars.

With distances available, in principle the whole universe could be mapped out in three dimensions. What is its structure? Newton thought that the stars may be infinite in number and distributed uniformly in space; each star would be at rest and remain so as it is pulled equally on all sides by the gravitational attraction of all the others. Perhaps he chose to ignore the most obvious large scale feature in the night sky, the Milky Way, which is comprised of a vast number of stars, or perhaps he was unaware of it because it is not very conspicuous in the northern hemisphere. In fact, in 1750 the English astronomer Thomas Wright (1711–1786) proposed that the Milky Way is a large disc of stars which we see edge-on because our solar system is embedded in it. But suppose that the universe is much, much bigger, and (on average) uniform on the very largest scales. The idea of an infinite number of unchanging stars in an eternal universe infinite in space and time runs into difficulties, as various people have noted over the centuries. Today, the credit for noting this paradox is given to Wilhelm Olbers (1758–1840). Olbers' Paradox states that the fact that the sky is dark at night rules out an infinite and eternal universe containing an infinite number of unchanging stars. The reason is that, in such a universe, the stars (which have finite sizes) would overlap in every line of sight, so the entire night sky would be as bright as the surface of the Sun in all directions. This paradox persisted until well into the twentieth century, when it was finally resolved by remarkable developments in cosmology.

It was Albert Einstein (1879–1955) who produced the mathematical and physical theory of space, time and matter which is still used today. After he completed his 1905 special theory of relativity (as described in a following section), Einstein had the grand ambition of broadening his theory into a general theory of relativity which would include acceleration and gravity; the special theory and Newton's laws would be special cases of this general theory. His main insight was to realize that there is no distinction between acceleration and gravity—they are equivalent. He needed advanced mathematics to create his general theory, and found them in the works of the nineteenth century mathematician Bernhard Riemann (1826–1866), who had developed the geometry of curved surfaces in multiple dimensions. In Einstein's general theory of relativity, space and time are treated as one 'spacetime' which interacts with matter. The presence of matter bends spacetime, so one talks of 'curved space'.

After brilliant and exhausting efforts (described in the next chapter) Einstein finally published his theory in 1916. He was understandably eager for his new theory to have some empirical foundation. It was realized in the 1800s that Newton's laws did not exactly explain the observed advance of the perihelion of Mercury; there was a discrepancy of 43 arcseconds per century. Various explanations were proposed, including a perturbation by some small unseen body or a smooth distribution of unknown 'dark matter'. As it turned out, no unknown matter was required; Einstein showed that his theory completely explains the discrepancy. He also proposed two other tests of his theory: the bending of light by the gravitational potential of the Sun, and the redshifting of light in a strong gravitational field.

It was known by the early nineteenth century that the bending of light by a massive object is predicted by Newton's theory of gravity. The corpuscles of light from a distant star just skimming by the surface of the Sun would be deflected by about 0.9 arcseconds. In his 1916 paper Einstein showed that general relativity predicts a deflection twice as great, and proposed that a measurement be made. In spite of the war raging at the time, this became known to the English astrophysicist Arthur Eddington (1882–1944), who proposed that the Royal Society make plans to observe an eclipse of the Sun that would occur in 1919. The idea was to first photograph the eclipse, showing the surrounding star field, and then (at night time, months later) photograph the same area of sky with the Sun out of the way. Comparing the two photographs should reveal that the stars closest to the disk of the Sun during the eclipse appear shifted in position outwards from the Sun. The shifts were found, at exactly the value Einstein had predicted. This made headline news around the world.

Einstein understandably wanted to apply his general theory to the largest of all entities—the universe—which he did in a 1917 paper. He knew that the current view amongst astronomers was that the universe is static. This was supported by the relatively small motions of even the most distant stars. However, according to his equations the universe could not be static—it had to be either expanding or contracting. But he then realized that the simple addition of a constant (now called the cosmological constant) to his equations could make the universe static. He included the constant and published his paper. He would realize later what a huge mistake this had been.

Meanwhile, important relevant work was progressing on observations of variable stars, which would play a key role in discoveries about the distant universe. Henrietta Swan Leavitt (1868–1921) was working in a Harvard team studying a class of stars called 'Cepheid variables' which had been observed in a 'cloud' of stars known as the Small Magellanic Cloud (SMC). Finding these stars on the many photographic plates was a painstaking process. All Cepheids go through regular cycles of brightening and dimming, with periods ranging from 1 to 70 days.

What Leavitt eventually found was that the brighter Cepheids have longer periods. By 1912 she had enough data on 25 Cepheids to express this period-brightness relationship in a mathematical formula. She realized that the reason this relationship shows up so clearly is that the SMC is so far away that all the stars are effectively at about the same distance from us. So the relationship is actually between the period and the *intrinsic luminosity*. This was a spectacular discovery, making it possible to measure cosmic distances. All that was needed was to find the distances to a few Cepheids in our local neighbourhood (using the parallax method), so that their intrinsic luminosities could be determined. Then the intrinsic luminosities of all other Cepheids could be deduced from the period-luminosity relationship. In this way it became possible to determine the distances of far-away Cepheids just from measurements of their periods and apparent brightnesses. We can measure the distances to far-away stars and galaxies!

Herzsprung was the first to measure the distances to nearby Cepheids in 1913. His calibration implied a distance to the SMC of 30,000 light-years! (We now know that the distance, corrected for extinction by dust, is actually 197,000 light-years).

This work on Cepheids came just in time to be used on data from the largest telescopes in the world, first the 1.5-m reflector on Mount Wilson, California, and then the 2.5-m Hooker telescope on the same mountain, completed in 1918. Harlow Shapley (1865–1972) was the first to measure the 'Milky Way Galaxy' with these telescopes using the Cepheids. It was widely thought that the Milky Way dominated the universe, and that the Sun was at the centre of this disk of stars, but he showed that the centre of the Milky Way is actually about 30,000 light-years away from us, and he estimated that the Milky Way itself is about 300,000 light-years across. There were other diffuse patches of light, but many thought that these were just satellites of the Milky Way Galaxy or stellar or gaseous nebulae within the Milky Way. Some others, most notably the American Heber Curtis (1872–1942), thought they may be distant galaxies in a vast universe of galaxies, of which our Milky Way Galaxy is just one.

Edwin Hubble (1889–1953) arrived at Mount Wilson just in time to make some of the greatest discoveries ever made about the universe. He studied mathematics and astronomy at the University of Chicago, and was one of the early Rhodes Scholars at Oxford. He also studied law, but did not pursue that career. At the age of 25 he started graduate studies in astronomy at the Yerkes Observatory of the University of Chicago, and received his PhD in 1917. He then joined the U.S. Army, but was too late to see combat. After a year in Cambridge, he moved to a staff position at the Mount Wilson Observatory where he remained for the rest of his life.

In 1925 Hubble was able to resolve individual stars in a spiral nebula (the Andromeda Nebula, now known as the Andromeda Galaxy). He identified several Cepheids in the nebula, and was able to determine its distance as about 900,000 light-years. This was a major discovery in its own right, as it showed that at least some of the spiral nebulae are in fact distant galaxies. He then found Cepheids in several other similar spiral nebulae, and again large distances were indicated. He pushed the limits of the telescope to find the most distant galaxies accessible at that time.

What was becoming clear was that there is a relationship between the distance to a galaxy and its redshift. Vesto Slipher (1875–1969) worked at the Lowell Observatory and could obtain redshifts, but not distances. What he did find was that 39 of his nebulae had redshifts, while only two had blueshifts. The natural interpretation of these results was that most of the nebulae are moving rapidly away from us. In 1927 the Belgian priest and astronomer Georges Lemaître (1894–1966) published a paper in which he derived and explained the relation between distance and the recession velocities of galaxies in terms of a universe expanding from an origin (for this he used Einstein's general theory of relativity, but dispensed with the assumption that the universe is static). But it was the superior results of Hubble in 1929 that best showed the simple fact that the redshift of a galaxy is proportional to its distance from us that made the most impact. It was a stunning result, and became known as Hubble's Law. The whole universe is expanding!

When Einstein heard about the expanding universe, he said that inserting the cosmological constant had been "the biggest blunder of my life", and he removed it from his equations in disgust in 1931. If he had left his equations in their original form, he would have been credited with predicting the expansion of the universe—a great discovery. But nevertheless his theory of general relativity did become the theoretical and mathematical framework for the expanding universe, and the much-maligned cosmological constant was destined to return in a later context at the end of the century.

A longstanding mystery over the centuries was how the stars shine. What is their prodigious energy source, and how can they keep going for so long? In the late nineteenth century there was a conflict between the huge timescales that seemed to be required for geology and evolution on the one hand, and the short lifetimes predicted for the stars by physicists on the other. William Thomson (Lord Kelvin) (1824–1907) assumed that the Sun's energy comes from its gravitational energy being converted into heat as it slowly contracts under its own weight, and estimated the Sun's age to be about 30 million years. But before the end of the nineteenth century physicists had discovered radioactivity, a possible source of the Sun's constant energy output, although the lifetimes still turned out to be far too short. Nevertheless, the idea that some sort of subatomic process might be involved was gaining traction.

Given the developments in subatomic physics and Einstein's famous equation $E = mc^2$ (showing the equivalence of mass and energy) in the early part of the twentieth century, Eddington was able to pronounce in 1920 that the subatomic energy in matter is nearly inexhaustible, and sufficient to supply the Sun for 15 billion years. The energy comes from the fact that the mass of a helium atom is less than the total mass of the four hydrogen atoms that combine to make it, so energy is released by nuclear fusion reactions 'burning' hydrogen into helium. An important realization came at the end of the 1920s, when Albrecht Unsold (1905–1995) and William McCrea (1904–1999) established the huge predominance of hydrogen atoms in the Sun compared to other elements.

The key interactions in the nuclear fusion process were identified by a number of physicists in the late 1930s, in particular Hans Bethe (1906–2005) and Carl von Weizsacker (1912–2007). These studies also benefitted from the wartime effort to develop nuclear weapons, and later, nuclear reactors. But even by the early 1950s it was thought that nothing could halt the central collapse of a star at the end of hydrogen burning. This view was completely changed in 1952 when Edwin Salpeter (1924–2008) showed how helium burning could proceed, leading to the creation of the 'heavier' elements (those with more protons in their nuclei, such as carbon, oxygen and neon). This breakthrough had major implications not only for stellar evolution, but also for cosmology and even life in the universe. But it was still not the complete story. Two years later Fred Hoyle (1915–2001) showed that Salpeter's mechanism would not be fast enough to give the observed ratios of the heavy elements, and he predicted a crucial 'resonance' in the nucleus of carbon-12 that was subsequently found at the predicted energy. The path was now clear for the creation of the heavy elements. Finally in 1957 the definitive review paper on how the elements are built up in stars was published by Margaret Burbidge (1919–), Geoffrey Burbidge (1925–2010), William Fowler (1911–1995) and Fred Hoyle.

So it was established how the stars shine, and that almost all of the elements are made in stars. The very atoms in our bodies were made in stars!

But this normal process of nucleosynthesis in ordinary stars only produces the elements up to iron. Elements heavier than iron (such as copper, silver, gold and uranium) had to be made in some other way. It was eventually found that the short but extremely energetic interactions produced in supernova explosions could produce these heavier elements in adequate abundances. Increasingly sophisticated computer models could follow these interactions in detail, but observational data were hard to come by, as supernovae are rare and most are distant. Fortunately, in 1987 a supernova explosion occurred in the Large Magellanic Cloud, a close neighbour to our galaxy; it was the closest supernova since Galileo's time. An abundance of detailed observational data was obtained using a wide range of the most powerful telescopes and satellites. Now even the supernovae were well understood.

Returning to Hubble's 1929 discovery, George Gamow (1904–1968) and his colleagues Ralph Alpher (1921–2007) and Robert Herman (1914–1997) took the expansion of the universe seriously in the late 1940s, and worked out the possible consequences of the early universe having been in a very dense and therefore very hot state (what Hoyle facetiously called a 'Big Bang'). One important consequence was that it would lead to the creation of the light elements in the universe—hydrogen, deuterium, helium and lithium. A paper announcing this important prediction was published by Alpher and Gamow in 1948,[9] and the predicted abundances were subsequently confirmed by observations.

Another consequence of an early hot phase would be the presence a decaying afterglow. The radiation would have cooled as the universe expanded, and would have been redshifted to much longer wavelengths. In 1948 Alpher and Herman computed that it would now be as cold as 5 K (5° above 'absolute zero', or −268 C), and detectable at millimetre wavelengths.[10]

In 1964 two radio astronomers, Arno Penzias (1933–) and Robert Wilson (1936–) were using an antenna that had originally been designed for satellite radio communications at Bell Laboratories to accurately measure the emission from radio sources and the sky background. They had to identify and remove all extraneous noise from their data, and, after much effort (they even delicately removed two pigeons and their droppings from their antenna), they were left with a persistent noise of 3.5 K, a hundred times more than expected and constant over the entire sky. At the same time, astrophysicists Robert Dicke (1916–1997), Jim Peebles (1935–) and David Wilkinson (1935–2002) were preparing to search for the afterglow of the Big Bang at Princeton University, just 40 km away. When Penzias was informed about the Princeton work he phoned Dicke, and they realized that Penzias and Wilson had serendipitously discovered the Cosmic Microwave Background (CMB), as it is called. It was a huge discovery, considered to be the definitive evidence in favour of the Big Bang model.

The CMB was subsequently observed by three major spacecraft (COBE, WMAP and Planck) and many ground-based and balloon experiments, and its properties were found to contain a treasure trove of information about the large-scale properties and origin of the universe. The CMB has a very special spectrum—it is the most perfect 'black-body' spectrum known to man, and this characteristic virtually confirms that it is the afterglow of the Big Bang. It is uniform over the sky to a precision of one part in 100,000. And yet it contains

---

[9] (Gamow added the name of Hans Bethe as a joke, to make the paper Alpher-Bethe-Gamow as in α-β-γ; neither Alpher nor Bethe were amused)

[10] Radio, millimetre, infrared, optical, ultraviolet, X-ray and gamma-ray emissions are all part of one vast and continuous 'electromagnetic spectrum'. What we normally call 'light' is just the very narrow optical range in the middle of the spectrum.

the faint imprints of embryonic galaxies, seen as they were just 380,000 years after the Big Bang.

But over the past century there arose two major mysteries about the universe that have yet to be resolved. In the 1930s Swiss astronomer Fritz Zwicky (1898–1974) was studying the motions of the outermost members of a cluster of galaxies, and found that they are moving far faster than expected from Newtonian physics: their motions should have ejected them from the cluster—but the cluster is clearly intact. Zwicky hypothesized that there might be a large amount of unseen matter in and around the cluster that provides sufficient gravitational mass to keep the cluster intact. Work by Vera Rubin (1928–2016) and others in the 1960s and 1970s found something similar in the speeds of stars and gas in the outer regions of galaxies. The outermost stars and gas could only be kept in place if the galaxies are surrounded by giant massive 'halos' of unseen 'dark matter'. Other lines of evidence pointed in the same direction. It is now clear that the total mass of dark matter in our universe is over five times the mass of the ordinary matter we know about. The ordinary matter is just 'froth' on a vast unseen ocean of dark matter. Many studies and searches have been made, indicating that the missing mass must be in some exotic form, such as elementary particles we do not yet know about. We are missing one of the major ingredients of the universe.

Now we also have 'dark energy'. In the 1990s two large groups of astronomers led by Saul Perlmutter (1959–), Brian Schmidt (1967–) and Adam Riess (1969–) were using distant supernovae to detect the long-sought-after deceleration parameter of the universe (all the matter in the universe causes the expansion to slow because of gravitational attraction). Instead, in 1997 they discovered that the expansion of the universe is *accelerating*. This has been attributed to a mysterious 'dark energy' which has a repulsive force across the universe (similar to the cosmological constant introduced by Einstein in 1916). Matter and energy are equivalent according to Einstein's $E = mc^2$; the dark energy, dark matter, and ordinary matter contribute 68%, 27% and 5% respectively to the total mass-energy of the universe. So we now find that we do not know what 95% of the universe is made of!

But in spite of the continuing mysteries of dark matter and dark energy, the large-scale properties of our universe have now been rather accurately measured, and it is said that we are living in the era of 'precision cosmology'. These properties have led to remarkable new hypotheses about the origin of our universe, and the possibility that there may be other 'universes'—perhaps an infinity of them. In 1979 Alan Guth (1947–) was puzzling over certain problems in the standard Big Bang model of the universe. He realized that they could all be resolved if our universe had experienced a brief period of hyper-expansion extremely early in its existence. This became a very popular hypothesis, which came to be known as 'inflationary cosmology'. From that it was a small step to suggest that this inflationary episode was not actually a *part* of the Big Bang universe, but that it was the *cause* of the Big Bang. This and other ideas have led to the concept of a 'multiverse', in which universes are being formed all the time due to spontaneous creation events in an infinite 'quantum vacuum'. Searches are currently being made for any evidence regarding these hypotheses.

Finally, we come back to where we started in this section: the solar system. As we know of one star surrounded by planets (our solar system), it is reasonable to ask whether there may be others. Finding planets orbiting other stars may seem an impossible task, as stars are so much brighter than planets. But, in one of the major astronomical discoveries in the last century, the first such 'extrasolar planet' orbiting a normal star was found in 1995 by Swiss astronomers Michel Mayor (1942–) and Didier Queloz (1966–) using a novel technique. Since then over 3700 extrasolar planets have been found; there are probably more than a hundred billion in our galaxy alone.

Almost all areas of astronomy have benefitted enormously from the many new technologies that have become available over the last century. The entire electromagnetic spectrum has become accessible for observations: the radio, millimetre and optical wavebands (observable with ground-based telescopes) and the infrared, ultraviolet, X-ray and gamma-ray wavebands (observable with telescopes in space). Giant telescopes have been built, outfitted with huge high-tech instruments backed up by massive computing power. A large number of discoveries have been made, including quasars, pulsars, radio galaxies, black holes, X-ray binaries, gamma-ray bursts, protostellar jets, interstellar masers, the still-mysterious 'Fast Radio Bursts' and many more. It has been a bonanza.

Aside from the electromagnetic spectrum itself, there are three other observational windows on the universe: cosmic rays (energetic particles that interact with the Earth's atmosphere), neutrinos (ghostly particles that can pass right through the Earth), and gravitational waves (distortions in spacetime itself).

Cosmic rays have been studied for many decades. They are almost all protons and charged atomic nuclei, they originate in the Sun and beyond the solar system, and they have provided unique information for fundamental physics. Unfortunately, however, the extrasolar cosmic rays are deflected from their original paths by galactic magnetic fields, so their sources cannot be accurately pinpointed. Nevertheless, last decade a collaboration of over 400 scientists from 69 institutes in 16 countries built the huge Pierre Auger Observatory in Argentina to study the high-energy cosmic rays (those above 10^{18} electron volts) by observing the 'air showers' they produce. The observatory consists of 1600 car-sized water-tank detectors distributed over 3000 km^2 together with 24 dedicated telescopes. The large size is required because the energetic cosmic rays are very rare—only one per km^2 per century at the highest energies. Recently, 12 years of observations recording tens of thousands of events were analysed, and from their overall distribution on the sky it was possible to conclude that these ultra-high energy particles are most likely extragalactic—they come from beyond our galaxy.

Neutrinos, unlike cosmic rays, can pass unimpeded through almost anything, so a sensitive neutrino telescope can identify their sources and study their physics. But precisely because they can pass through almost anything they are extremely difficult to detect. The Sun produces a huge flux of neutrinos, and these have been detected in several experiments. For some decades there seemed to be only half the number expected from theory (the 'Solar Neutrino Problem'), but this discrepancy has now been resolved. The 1987 supernova in the Large Magellanic Cloud mentioned above produced a huge number of neutrinos (about 10^{58}), and 19 of them were detected by two experiments in deep mines in the U.S. and Japan. These few detections produced a wealth of information about neutrinos, and marked the beginning of neutrino astronomy. Ever larger neutrino telescopes are being developed, as described in Chap. 4.

Recently a totally different window on the universe was opened up with the discovery of gravitational waves—the distortion of spacetime itself by a distant cataclysmic event. This type of radiation had been predicted by Einstein in 1916 in his general theory of relativity, but even he doubted that it would ever be detected due to the weakness of the signals. But since the 1960s increasingly sophisticated efforts have been made. The Laser Interferometer Gravitational-Wave Observatory (LIGO), comprised of two large interferometers on opposite sides of the United States, finally made the breakthrough detection in September 2015. A very specific and complex 'chirp' signal caused by the merger of two distant supermassive black holes (29 and 36 times the mass of the Sun) was simultaneously detected by the two interferometers, in spectacular agreement with the predictions of Einstein's theory.

A few more such events were detected over the next couple of years by LIGO and a European counterpart called Virgo; the most interesting of these was a merger of two neutron stars in August 2017. The merger of supermassive black holes leaves no debris, but the merger of less massive neutron stars produces relativistic jets, nuclear reactions and an abundance of hot gas—a feast of electromagnetic radiation for our more conventional observatories. NASA's Fermi Gamma-Ray Space Telescope detected a gamma-ray flash less than two seconds after the neutron-star merger, the new point of light in the sky was quickly identified by ground-based telescopes, and over 70 teams of observers on all seven continents worked around the clock in a frenzy to study the aftermath of this 'kilonova'. It was a rare opportunity. It produced what was probably the greatest mobilization of astronomers ever and a blizzard of papers, one of which has thousands of co-authors. At a distance of only 130 million light-years this merger was the closest event (so far) for both gravitational-wave and gamma-ray astronomy. While it has provided a huge amount of new information on many astrophysical processes, it has also added further confirmation for Einstein's theory, which correctly predicted both the complex gravitational-wave signal and the temporal coincidence with the gamma-ray burst (gravitational waves also travel at the speed of light). More gravitational-wave interferometers are being built around the world and in space, and there is no question that gravitational waves are already providing an important new window on the universe.

2.9 The Very Small

What is the smallest entity? The ancient Greeks called it the atom. The earliest thinkers we know of who pondered this question were the Greek philosophers Leucippus, Democritus and later Epicurus, as mentioned earlier. They believed that all matter is composed of atoms separated by empty space. In contrast, Aristotle specifically rejected atomism because he opposed the idea of a vacuum or a void, whereas atomism envisaged atoms moving about in a void and interacting with one another. The next mention of this question was made some two thousand years later, at the time of the Scientific Revolution.

Descartes rejected the idea of atoms for the same reason as Aristotle: "Nature abhors a vacuum"—the denser material would immediately fill a void. On the other hand, the revival of atomism was Pierre Gassendi's (1592–1655) most important contribution to science. He thought that there was nothing between the atoms, and that they could join together to form what he called molecules. There was in fact evidence that a void could exist. Evangelista Torricelli (1608–1647) was an Italian scientist who knew Galileo late in his life. Stimulated by some ideas from Galileo, he invented the first barometer in 1643. He filled a glass tube, sealed at one end, with mercury, and inserted it open-end down into a basin of mercury. The column of mercury then fell, leaving a gap between the top of the column of mercury and the top of the glass tube. Descartes knew about this experiment, but still insisted that there is a much finer fluid which fills all the gaps and prevents the existence of a vacuum, even in deep space.

After Descartes' death, Otto von Guericke (1602–1686) invented a vacuum pump so perfect that it could extinguish candles and silence the ringing of a bell as the air was removed. In 1657 he made two half-metre diameter hemispheres and pumped all the air out of them, locking them together with a vacuum seal. The vacuum was so good that 16 horses, eight harnessed to each side of the globe, could not pull the halves apart. Descartes would undoubtedly still have insisted that his super-fine fluid is everywhere, but for Newton, on the other hand, the void in empty space is absolute.

Robert Boyle (1627–1691) carried these ideas further. In his most famous experiment, he used a glass tube in the shape of a J, with the top open and the short end closed. He poured mercury into the tube to fill the U bend at the bottom, sealing off the air in the short end. He could then increase the pressure on the air at will by pouring more mercury into the long end. What he found became known as Boyle's law: the volume occupied by a gas is inversely proportional to the pressure on it. He noted that this can easily be explained by an atomic concept, but not by Descartes' model of the world. He rejected the four ancient 'elements'—earth, water, air and fire—on experimental grounds. He practiced alchemy, but tried to bring the scientific method into it, and he wrote a famous book *The Sceptical Chymist,* which became a turning point as alchemy evolved into chemistry. He supported the atomic hypothesis, suggesting that atoms can move freely in liquids, but are joined together in various ways, depending on their shapes, to make different solids. The role of chemistry, in his view, was to determine what things are made of.

In 1691 Halley tried to estimate the size of atoms using the gilding of thin wires. He asked craftsmen how much gold they used in drawing and gilding silver wire. From that information, and from the diameter and length of the wire, he estimated that the thickness of the gold around the silver was 120 nm—an upper limit on the size of an atom of gold. We now know that the actual size is a thousand times less than that, but at least this first attempt had been made.

Ironically, the road to the very very small did not involve the microscope, which was invented in the early seventeenth century. Instead, it involved the study of gases, which led to chemistry, molecules, atoms and even smaller entities.

Joseph Black (1728–1799) was a young man working on his doctorate at the University of Edinburgh when he did the research that made him famous. To make his important discoveries possible, he developed a sensitive balance which was far more accurate than any other at that time. He found that limestone could be heated or treated with acids to produce a gas he called 'fixed air', which was denser than air and extinguished both fire and animal life. We call it carbon dioxide. He showed that air is a mixture of gases, which was revolutionary at the time.

Another ten gases, including ammonia, nitrous oxide and carbon monoxide, were identified by Joseph Priestly (1733–1804). His most famous discovery was oxygen. In all this work he interpreted the results in terms of the infamous phlogiston theory promoted by the German chemist Georg Stahl (1659–1734). According to this theory, burning is explained by a substance (phlogiston) leaving the material that is being burnt. Priestly did not make the connection between burning and oxygen, but he did observe some of the characteristic properties of this new gas—that a lighted candle would flare up when being immersed into the gas, and that a mouse thrived in a sealed vessel filled with the gas.

Henry Cavendish (1731–1810) came from an aristocratic family in England. He attended a private school and then Cambridge. He was introduced to the world of science by his father, and he devoted himself to it for the rest of his life. His researches covered chemistry, physics and the Earth. He was noted for the precision of his work. He was also noted for his discovery of hydrogen, or what he called 'inflammable air', and he made careful experiments on its properties. He showed that water is not an element but rather a mixture of two other substances—an important clue for future work.

The man who pulled all these discoveries together into a real science of chemistry was Antoine Lavoisier (1743–1794). His father was a lawyer in Paris, and Lavoisier studied law at the University of Paris, but he also took courses in mathematics and science, and they determined the direction of his career. He was independently wealthy, so he could follow his whim. In addition to developing his scientific interests, he became a member of several aristocratic councils and an administrator of the hated *Ferme Generale* (a 'tax farm'). These helped to fund his scientific research, but they also led to his death: he was guillotined in 1794. But his scientific career was stellar. He is widely considered the 'father of modern chemistry'. He is known for recognizing the role oxygen plays in combustion, and he put an end to the phlogiston theory. He produced the first extensive table of elements. The book he published in 1789, *Traité Élémentaire de Chimie (Elementary Treatise on Chemistry)*, is sometimes regarded as chemistry's equivalent to Newton's *Principia*.

Humphry Davy (1778–1829) became a prominent scientist with no formal education beyond a provincial grammar school. He was the son of a Cornish farmer eking out an existence. He managed to teach himself French, and read Lavoisier's *Traité Élémentaire* in the original language when he was just 18. In the following year he became an assistant at a new research institute in Bristol. He carried out experiments with nitrous oxide; this became widely known as 'laughing gas' because of its intoxicating properties, and it made his name. He became a lecturer at the newly established Royal Institution in London, and within a year, at the age of 23, he was appointed professor of chemistry. In the course of his work he isolated potassium, sodium and chlorine, and he became President of the Royal Society. It was an amazing transition from a self-taught amateur to the highest post in British science.

Another scientist who rose to prominence from humble beginnings was John Dalton (1766–1844). His father was a weaver, and Dalton attended a local Quaker school. When he was just 15 he joined his brother and cousin in running a Quaker school, and he stayed there until he was 27. In addition to his duties at the school, he started to give public lectures. He became well enough known to be offered a teaching position in Manchester, and a few years later he found he was able to make a living as a private tutor; this gave him time to do science, and he stayed in Manchester for the rest of his life. One interest he had was the nature of mixtures of gases, culminating in the law of partial pressure, known as Dalton's Law. He also contemplated what the elements are made of, and came up with an atomic theory. In his view, each element is made up of indestructible atoms unique to that element, which can combine with atoms of other elements in simple ratios to form chemical compounds. The atoms of the different elements are differentiated by their size and weight. Dalton published these ideas in his book *A New System of Chemical Philosophy* in 1808; it included a list of estimated 'atomic weights'. His theory had a mixed reception, as some (again) found it hard to accept the idea of atoms separated by empty space, but it slowly gained acceptance as a heuristic model of reality. Dalton was awarded many honours and became a Fellow of the Royal Society.

The Swedish chemist Jöns Berzelius (1779–1848) was raised by an uncle after both his parents died when he was young. He studied medicine at the University of Uppsala, and then moved to the College of Medicine in Stockholm. He became a professor there at the age of 28, and his interests shifted to chemistry. He studied the proportions of elements in a large number of chemical compounds, and compiled a table of relative atomic weights for all 40 elements known at that time. His work provided strong evidence in favour of Dalton's atomic theory.

Two other developments in the early 1800s advanced the atomic theory. First, the French chemist Joseph Louis Gay-Lussac (1778–1850) found that gases combine in simple proportions by volume (water vapour is two parts hydrogen and one part oxygen by volume). Stimulated by this finding, Italian physicist Amadeo Avogadro (1776–1856) hypothesized that a given volume of any gas at a given pressure and temperature contains the same number of particles (atoms or molecules).

The idea of the atom (and molecule) was becoming common currency later in the nineteenth century, thanks to work by Edward Frankland (1825–1899), Archibald Couper (1831–1892) and Friedrich Kekule (1829–1896), among others. The term valency was introduced—the ability of one element (or atom) to combine with another. And the concept of bonds between atoms was easy to visualize, such as the linking of carbon atoms into rings with bonds linking them with other atoms.

In the 1860s four scientists independently 'discovered' the famous periodic table of the elements. They realized that if the elements are arranged in order of their atomic weights, there is a periodic pattern in which elements separated by multiples of eight times the atomic weight of hydrogen have similar chemical properties. The first three were the French minerologist Alexandre Beguyer de Chancourtois (1820–1886), the English chemist John Newlands (1837–1898), and the German chemist Lothar Meyer (1830–1895). All of them presented the essence of the periodic table, but none was as compelling as that of Dmitri Mendeleev (1834–1907).

Mendeleev was born in Siberia, the youngest of 14 children. His father, a schoolmaster, became blind when Mendeleev was very young; his mother supported the family with a glass works which were destroyed by fire when Mendeleev was 14, a year after his father died. His mother then took him to St Petersburg to receive a proper education. Unable to obtain a university place, he became a teacher. He was eventually able to obtain a master's degree in chemistry at the University of Saint Petersburg, where he worked for 2 years before going on a government-sponsored programme to Paris and Heidelberg. Returning to Russia, he completed a PhD and became professor of chemistry at the University of Saint Petersburg, where he remained for many years.

In 1869 Mendeleev published his famous paper *'On the Relation of the Properties to the Atomic Weights of Elements'*. He went further than just noting a broad pattern. Like the others he arranged the elements in rows of eight, in order of increasing atomic weight; the elements of similar chemical properties were then aligned in the columns of the table. But there were some irregularities and gaps. He took the initiative of changing the order of some elements which had similar atomic weights to make sure that all columns contained only elements of similar properties. He also left three gaps in the table, claiming that these correspond to three as yet undiscovered elements, and he predicted what properties these elements would have. And indeed, by 1886 all three of these elements had been discovered, giving a huge boost of confidence in the periodic table. There was no question that a fundamental property of the chemical world had been found.

The quest for the most fundamental constituents of matter was now passing hands from the chemists to the physicists. In the second half of the nineteenth century physicists were developing the kinetic theory of gases, which is based on the motions of the constituent atoms and molecules (assuming that they exist). Two of the leaders in this field were the Scot James Clerk Maxwell (1831–1879) and the Austrian Ludwig Boltzmann (1844–1906). In a paper presented in 1859, Maxwell showed that molecules in air at 15 °C experience more than eight billion collisions per second, with a mean free path of six millionths of a centimetre. This frenzy of activity appears to us on the macroscopic scale as a smooth, continuous gas. Maxwell showed the relationship between heat and motion: the temperature is a measure of the mean speed of the molecules. Boltzmann further developed this theory, and the distribution of the velocities of the molecules about their mean became known as the Maxwell-Boltzmann distribution.

Various estimates were by then being made for the sizes of the molecules. In 1865 the Austrian Johann Loschmidt (1821–1895) reasoned that in a liquid the molecules are touching each other but in gaseous form they are separated, and used the mean free path to estimate the size of the molecules that make up the air to be about a millionth of a millimetre, not far from the modern value.

One of the four famous papers that Einstein published in 1905, his *annus mirabilis*, concerned Brownian motion. It may seem a rather obscure topic for Einstein to have taken up, but he was determined to prove that atoms are real. In 1827 the Scot Robert Brown (1773–1858) had noticed that pollen grains floating in water, as seen through a microscope, move in a very erratic fashion. It was soon established that this kind of movement is present in any tiny grains suspended in a liquid or in air, and the phenomenon was called Brownian motion. It was suggested that this could be due to impacts on the grains by the molecules of the medium, but there seemed to be no way of proving this hypothesis. Einstein's paper finally gave a clear statistical description of the phenomenon. The grains follow a 'random walk', in which the distance traveled is proportional to the square root of the time taken. Einstein predicted a rate of six thousandths of a millimetre per minute for micrometre-sized particles in water at 17 C. The French physicist who took up the challenge of measuring this motion, Jean Perrin (1870–1942), was able to confirm Einstein's prediction. The physical reality of atoms and molecules was finally proven.

But is the atom really the smallest entity? Or does it have substructure? Are there even smaller entities? In the 1860s studies made using vacuum tubes with an electric current flowing between the positive and negative electrodes revealed that the glowing rays emitted from the cathode (negative electrode) seemed to follow straight lines. These became known as cathode rays. It was suggested that they may be composed of particles of matter. William Crookes (1832–1919) showed that a sharply defined shadow resulted if an obstruction was placed in the tube, and that the rays could cause a tiny paddle wheel to rotate—an indication that they carry momentum. In 1894 J.J. Thomson (1856–1940) showed that they move much more slowly than light, and by 1897 there was increasing evidence that they carry electric charge, are deflected by magnetic fields, and cause an intervening metal plate to become negatively charged. In 1897, having measured the charge-to-mass ratio of the cathode rays, Thomson concluded that the particles comprising the cathode rays are more than a thousand times smaller than an atom—he had discovered the electron, the first known subatomic particle.

At about the same time, the German theoretical physicist Max Planck (1858–1947) was working on a very different problem: black-body radiation. A perfect black body is one that absorbs all the radiation that falls on it, and for which the emitted radiation depends only on its temperature. The spectrum of the emitted radiation had been measured in many experiments, but it proved extremely difficult to make a mathematical model that would produce such a spectrum in its entirety. Planck finally succeeded in 1900, but at a significant cost: he had to assume that the energies of the hypothetical radiating 'oscillators' are multiples of an elementary unit—they are 'quantized'. Initially he considered this to be just a heuristic assumption—but it worked. This assumption, incompatible with classical physics, marked the birth of quantum physics.

This brings us to another of Einstein's four famous papers of 1905: *On a Heuristic Viewpoint Concerning the Production and Transformation of Light*. In it, he proposed that light itself is quantized: it behaves as if it consists of mutually independent quanta, or photons. In the same paper he applied this hypothesis to the photoelectric effect, in which electromagnetic radiation can knock electrons out of the surface of a sheet of metal. So, is light a wave, as had been assumed over the past century, or a stream of particles, as in Einstein's new theory? The reality of the photoelectric effect was eventually proven by the

American Robert Millikan (1868–1953), who had actually set out to disprove the theory because he believed as many did that light is a wave.

Following his discovery of electrons, J.J. Thomson proposed a model for the internal structure of the atom in 1904, his 'plum pudding model'. He suggested that the atom is composed of electrons immersed in a 'soup' of positive charge which balances the negative charges of the electrons. This model was tested experimentally by a New Zealander, Ernest Rutherford (1871–1937), and his team at the University of Manchester by firing the newly discovered 'alpha particles' into a thin gold foil. By looking for deflections of the positively charged alpha particles, they were able to probe the interior of the gold atoms. They found that, while most of the alpha particles went straight through, a few were deflected, and to their astonishment they even found rare cases in which the alpha particles came straight back. Rutherford's conclusion in 1911 was that the atom contains a very small positively-charged nucleus, which deflects and reflects the positively-charged alpha-particles, surrounded by low-mass electrons. The nucleus only occupies about one hundred-thousandth of the diameter of an atom: atoms are mostly empty space containing electromagnetic fields.

The problem with Rutherford's model was that it was unstable—there was nothing to stop the negatively-charged electrons from falling into the positively-charged nucleus. One could imagine that the electrons are in orbits around the nucleus, like the planets around the Sun, but the orbiting electrons would quickly lose their energy by electromagnetic radiation and fall into the nucleus. The Danish physicist Niels Bohr (1885–1962) visited Rutherford in Manchester for 6 months in 1912, and it was there that he conceived his quantum model of the atom, published in 1913. He proposed that the electrons revolve around the nucleus in stable, discrete energy levels ('orbits'), but can jump from one energy level to another, in the process emitting (or absorbing) a discrete quantum of electromagnetic energy. In this way he was able to explain the observed spectral lines of hydrogen atoms (classical physics had no way of producing these discrete features). It was a remarkable achievement, and a major step for quantum physics. His model, with refinements made over the following decade, provided the basis for our understanding of chemistry.

In the 1920s a number of important papers on quantum physics were published. In 1924 Louis de Broglie (1892–1987) completed his PhD thesis at the Sorbonne in which he suggested that, just as electromagnetic waves can be described in terms of particles, so can all material particles, such as electrons, be described in terms of waves. *Everything* has dual wave-particle nature. This became known as wave-particle duality, a core theme in quantum mechanics. Not long after de Broglie made his bold hypothesis, electron diffraction was observed in two independent experiments, one by Clinton Davisson (1881–1958) and lester Germer (1896–1971) and the other by George Thomson (1892–1975). The astonishing reality of wave-particle duality was confirmed.

Two other conceptually different mathematical models describing the behaviour of electrons in atoms were published in 1926. One of them, by Erwin Schrödinger (1887–1961), was based entirely on waves, describing the subatomic world in terms of a wave equation. The other, by Werner Heisenberg (1901–1976), took the particle approach, involving 'quantum-jumping' between energy levels. Paul Dirac (1902–1984) produced a more abstract formalism, and showed that the other two approaches were contained within that formalism and were mathematically equivalent to each other. Very different theories, all correct, describing the same phenomena.

The most startling paper was that published by Heisenberg in 1927 on the famous 'uncertainty principle'. He showed that, according to quantum mechanics, it is impossible to simultaneously know all the properties of a system with perfect accuracy, even in principle. For example, if the position of a particle is known accurately, its momentum cannot be; the more accurate the position, the less accurate the momentum, and vice versa. The same applies to time and energy. This is not due to the limitations of our measurements—it is a fundamental property of nature itself. The values of these properties are determined by 'probability waves'. Einstein was always upset by this, and thought that someday an underlying reality would be found in which strict causality reigned rather than mere probability and uncertainty. He said "God does not play dice." But subsequent experiments made decades later refuted Einstein's view and confirmed the predictions of quantum mechanics. Bizarre as it may seem, that appears to be the way the world is.

Also in 1927, Dirac published what was regarded as the definitive wave equation of the electron. It had two solutions, the second of which seemed to describe a particle which has the same mass as an electron but a *positive* charge. It became realized that energetic photons, for example, could be converted into pairs of particles—ordinary negatively charged electrons and positively charged electrons. This phenomenon was subsequently observed in 1932 by Carl Anderson (1905–1991) in his studies of cosmic rays. He called the positive electron the positron. It was the first known particle of 'antimatter'; every particle is now known to have an 'antiparticle' with opposite charge.

Meanwhile, work continued on the properties of the atomic nucleus. Following up on studies made in Germany and France by bombarding nuclei with alpha particles, James Chadwick (1891–1974) made similar experiments in 1932 and concluded that the alpha particles were knocking previously unknown neutral particles out of the nuclei. He determined their mass to be slightly greater than that of the proton—he had discovered the neutron.

But what kept the nucleus together? Theoretical work by Wolfgang Pauli (1900–1958) and Enrico Fermi (1901–1954) showed that a short-range 'strong nuclear force' could keep the protons and neutrons together, and that another short-range force, the 'weak nuclear force', could explain the process of radioactive beta decay, in which protons and neutrons can be converted into one another. The particles and forces of the atomic nucleus were finally being identified.

Subsequent theoretical work in the 1930s and 1940s ultimately produced a complete theory describing how electromagnetic radiation and matter interact, with full agreement between quantum mechanics and Einstein's special theory of relativity (mentioned below). Quantum electrodynamics (QED), as the theory is known, was developed independently by the Japanese Sin-itiro Tomonaga (1906–1979) and the Americans Julian Schwinger (1918–1994) and Richard Feynman (1918–1988). All their approaches are mathematically equivalent. A measure of the success of QED is given by the fact that its prediction for the 'magnetic moment' of the electron agrees with experiment to within one part in ten billion!

During the 1950s and 1960s a bewildering flurry of new subatomic particles was being discovered in the new particle detectors. A high-energy collision of two protons could result in a torrent of hundreds of new and unexpected particles, most of them very short-lived. In 1964 a major simplification to this 'hadron zoo' was provided by the quark model conceived by Murray Gell-Mann (1929–) and George Zweig (1937–). According to this model protons and neutrons contain three even more fundamental entities called 'quarks', which can never escape their parent protons and neutrons. These were first 'seen' in experiments in the late 1960s, which revealed them to be point-like, matching the quark model.

So—first atoms, then electrons, protons and neutrons, and now quarks—fundamental physics was probing to ever smaller scales, to 10^{-18} m and even smaller—millions of times smaller than atoms!

There was a continuing series of important theoretical and experimental developments in those heady days. In 1967 the 'Standard Model of Particle Physics' was initiated by Sheldon Glashow (1932–), Steven Weinberg (1933–), and Abdus Salam (1926–1996), unifying the electromagnetic and weak forces into an 'electroweak' force. Quantum Chromodynamics (QCD) is the modern version of nuclear physics that applies on the scale of quarks, and was an improvement on the quark model of Gell-Mann. David Gross (1941–), David Politzer (1949–) and Frank Wilczek (1951–) all made major contributions to QCD.

In 1984 large collaborations at CERN (the European Organization for Nuclear Research) led by Carlo Rubbia (1934) and Simon Van der Meer (1925–2011) discovered the massive carriers of the weak force. Most recently, in 2012, the Large Hadron Collider (LHC) at CERN discovered the 'Higgs boson', which is responsible for giving rise to the masses of all the fundamental particles. The 'Higgs mechanism' had been proposed almost 50 years previously by Peter Higgs (1929–) and a few other theorists. The Higgs boson was the last remaining piece of the Standard Model, and its discovery was a momentous event for fundamental physics.

The Standard Model of particle physics now classifies all the known subatomic particles, and includes the electromagnetic, weak and strong forces. It explains all of the experimental results from the world's accelerators. According to the Standard Model, the fundamental constituents of matter are the families of 'fermions' ('leptons' and 'quarks') and their antiparticles. The 'bosons' are the carriers of the fundamental forces: the 'photons' carry the electromagnetic force, the 'gluons' carry the strong force, and the 'W' and 'Z' particles carry the weak force. The Higgs boson gives rise to the masses of the particles.

The Standard Model provides an accurate physical foundation underpinning chemistry, biology, electronics, engineering, materials science, astrophysics, much of cosmology, and the physics of everyday life. QCD runs the world of protons, neutrons and other 'hadrons'; QED runs the world of light, atoms and chemistry. The Standard Model was a monumental *tour de force*—the most successful theory known to man—with predictions that agree with experiments to accuracies as good as one part in ten billion.

This is an extraordinary time. More and more LHC data are pouring in and being scrutinized in every detail, and they still agree with the predictions of the Standard Model, developed 40 years ago. One might conclude that the theory has been so successful that the job is now done—that the era of discovery is over, as far as the study of 'The Very Small' is concerned. But, never satisfied, physicists are anxiously looking for any possible 'new physics' beyond the Standard Model (which, in spite of all its success, is still considered to be incomplete and unsatisfactory in several respects). There are a number of possibilities they are exploring, and these are mentioned in the last chapter.

2.10 Light

What is light? The studies of optics, magnetism and electricity all date back to the ancient Greeks. Euclid had the insight that light travels in straight lines, and he studied the reflection of light. Thales explored magnetism, and he rubbed amber with a dry cloth to attract feathers (the Greek word for amber is *electra*). Is there any connection between these three phenomena?

They were all studied over and over again throughout the Greek Miracle, the Islamic and medieval periods, and the Scientific Revolution. By the early eighteenth century, as outlined above, Newton had established that white light is a combination of all the colours in a spectrum, it had been shown that light travels at a finite speed, and there were two competing theories about the nature of light—a stream of particles (Newton's corpuscular theory) or a wave phenomenon (as proposed by Christiaan Huygens, Robert Hooke and others). The corpuscular theory held sway for a century.

The next major step was taken by the Englishman Thomas Young (1773–1829). His name is almost synonymous with the famous 'double-slit experiment', in which he showed that two light beams, formed by passing light through two thin parallel slits, are diffracted and then interfere with each other in such a way that they enhance each other in some directions and cancel each other in other directions. The result is clearly a wave pattern, similar to that seen when the ripples from two pebbles dropped into a pond of water interact with each other. As he said to the Royal Society in 1803, "The experiments… may be repeated with great ease, whenever the Sun shines, and without any other apparatus than is at hand to every one." In spite of the clarity and power of this experiment, there was initial reluctance to give up the corpuscular theory of a giant such as Newton.

Not much later, a young French man by the name of Augustin-Jean Fresnel (1788–1827), a civil engineer working on road projects, developed an interest in optics as a hobby, unaware of Young's work. He eventually developed his own wave theory of light, and in 1818 submitted it in a competition to explain the properties of light, organized by the French Academy of Sciences. The three judges all supported Newton's corpuscular theory, and one of them, the mathematician and physicist Siméon Denis Poisson (1781–1840), thought he had found a fatal flaw in Fresnel's theory. He calculated that, according to Fresnel's theory involving diffraction, a beam of light should produce a bright spot exactly behind a small obstructing disc, where obviously (according to Newton's corpuscular theory and common sense) there should be the darkest shadow. The chairman of the committee, physicist Dominique-François-Jean Arago (1786–1853), performed the experiment, and, to great astonishment, observed the predicted bright spot. This convinced most scientists of the wave nature of light, a theory that then dominated throughout the nineteenth century.

Meanwhile, the phenomenon of electricity was also generating increasing interest in the eighteenth and early nineteenth centuries. The American Benjamin Franklin (1706–1790) made some important contributions to the study of electricity in the mid-1700s, for which he was elected a Fellow of the Royal Society. He is most famous for the idea of showing that lightning is electricity by flying a kite in a lightning storm, and he invented the lightning rod.

A huge contribution for subsequent studies of electric phenomena came from the Italian Alessandro Volta (1745–1827). Stimulated in part by a dispute with another Italian, Luigi Galvani (1737–1798), concerning the twitching of frogs' legs (which Galvani called 'animal electricity'), Volta realized that the frogs' legs served merely as conductors between two metals. Following a series of experiments he invented the 'Voltaic pile', known today as a battery. His invention was a pile of alternating silver and zinc plates separated by cardboard soaked in brine; it produced a steady flow of electricity when the top and bottom plates were connected by a wire. This momentous discovery was announced to a meeting of the Royal Society in 1800. Suddenly scientists could work with continuous electric currents that could be switched on and off as desired. Volta's battery became an indispensable tool in scientific research, as it is in everyday life today.

In 1820 the Dane Hans Christian Ørsted (1777–1851) noticed that a compass needle was deflected when an electric current in a nearby wire was switched on and off—a bizarre case of 'action at a distance'. The needle was caused to point at right angles to the wire. This was the first indication of a connection between electricity and magnetism.

Michael Faraday (1791–1867) became famous for his work on these topics. He was raised in poverty, but eventually became an apprentice to a bookbinder in London, which gave him the opportunity to read extensively. He also carried out experiments in chemistry and electricity, and was an active member of the City Philosophical Society. By persistence and a stroke of luck, in 1813 he became the laboratory assistant at the Royal Institution, working with none other than Humphry Davy on his chemical experiments. He eventually became known in his own right for his talents.

When he heard of Ørsted's experiments he devised his own in 1821, and found that he could make a wire carrying an electric current rotate continuously around a fixed magnet. This discovery, which led to the electric motor, made him well-known throughout Europe. Electricity can produce motion!

Faraday went back to his work on chemistry and became director of the RI laboratory and Fullerton professor of chemistry, but in 1831 he returned to electricity and magnetism, and made another momentous discovery. He found that if he moved a magnet through a loop of wire, an electric current was induced in the wire. Motion can produce electricity!

His 1821 and 1831 discoveries were beautifully symmetrical. In addition to the first electric motor he had now produced the first electric generator.

The story didn't end there. Faraday was cautiously probing some new speculations, according to which magnetic and electric 'lines of force' exist (we would now call them fields), that disturbances in them take time to propagate, and that light itself may be explained in terms of vibrations of the lines of force.

James Clerk Maxwell picked up where Faraday had left off. He was born in Edinburgh into a relatively prominent family, and raised in Galloway (south-western Scotland). His father was a lawyer, with a keen side-interest in science and technology. Maxwell's mother died when he was eight, and he had a rather brutal tutor, until he was sent at the age of ten to the Academy of Edinburgh. His father took him to see scientific demonstrations and a meeting of the Edinburgh Royal Society, and by the age of 15 his first scientific paper was published in the proceedings of that society. He studied at the University of Edinburgh and then Cambridge, where he became a Fellow of Trinity (Newton's old college). From there he became a professor in Aberdeen, and then in 1860 at King's College in London, where he produced his theory of electromagnetism which unified electricity and magnetism.

In 1864 he published his monumental paper *A Dynamical Theory of the Electromagnetic Field.* His equations showed that the speed of an electromagnetic wave is equal to $1/(\varepsilon_o\mu_o)^{1/2}$, where ε_o and μ_o are the electric and magnetic constants respectively. When he inserted the values of these two constants, as determined *independently* from electric and magnetic measurements, he discovered that the speed is exactly the same as that of light (to within experimental errors)! He wrote "We can scarcely avoid the conclusion that light consists in the transverse undulations of the same medium which is the cause of electric and magnetic phenomena". He had discovered that light itself is an electromagnetic phenomenon. Light, electricity and magnetism are all manifestations of a single physical phenomenon: electromagnetism. Maxwell's theory was the greatest step in physics since Newton's *Principia.*

But there was still a problem. If electromagnetism is a wave phenomenon, then waves in *what*? What is the medium? The concept of an *aether* permeating space goes back—of course—to the ancient Greeks (it means 'fresh air' or 'clear sky', and there was a Greek god, *Aether*). It was adopted in the seventeenth and eighteenth centuries as the name of the medium that carries light waves, and in the late nineteenth century for the medium that carries electromagnetic waves. If there is such a medium, we may be able to detect an 'aether wind' using the motion of the Earth going around the Sun. The Americans Albert Michelson (1852–1931) and Edward Morley (1838–1932) conducted a precise experiment in 1887 comparing the speed of light in different directions, showing that there is no detectable difference between the speed of light in the direction of the Earth's motion and at right angles to it, or in any other direction. This was the first significant evidence against the presence of an aether (and this very meticulous experiment has been facetiously called "the most famous 'failed' experiment in history").

In 1905 Albert Einstein took a totally different approach. He assumed that the constant in Maxwell's theory—the speed of light—really *is* constant, regardless of the motion of the observer, whether moving towards or away from the source of the light, or in any other direction. It was a totally non-intuitive idea, but he followed it to its logical conclusion. In one of his 'thought experiments', he wondered what it would be like to 'ride on a beam of light', and his tram rides to work in Bern made him think of how different observers perceive their motion and events around them. The result was his revolutionary special theory of relativity, published in the third of his four famous 1905 papers, entitled *On the Electrodynamics of Moving Bodies.* The implications were bizarre (such as the famous 'twins paradox' in which a space-traveling twin returns home and finds that he has aged much less than his stay-at-home twin), but they were unavoidable. Einstein's questions about light had led him to a complete revolution in our concepts of space and time. Gone were Newton's concepts of absolute space and simultaneity (although Newtonian physics continues to work as well as always for everyday life). To cap it all off, at the end of 1905 Einstein wrote a three-page 'spinoff' paper following from his special theory of relativity: *Does the Inertia of a Body Depend Upon Its Energy Content?* There he introduced his famous equation $E = mc^2$.

By this time the proposed nature of light had gone through several iterations, including Euclid's rays, Huygens' and Hooke's waves, Newton's corpuscles, Young's and Fresnel's waves, and then, from Einstein's other famous 1905 paper on the photoelectric effect mentioned in the previous section, once again 'particles' (in this case photons), as confirmed by Millikan. But even that wasn't the end. In the 1920s light (like everything else) was seen to be described as both particles *and* waves. Eventually it was subsumed into the Standard Model of Particle Physics; the electromagnetic force was unified with the weak force into the 'electroweak force', and still further unification is foreseen.

So light has certainly threaded a rather tortured path throughout the entire history of science.

2.11 Life Itself

What is the basis of life? As with the questions raised in the previous sections, the study of life also goes back to the time of the ancient Greeks. Aristotle was considered to be the 'father of zoology', as he studied and classified hundreds of species, and Theophrastus (c. 371–287 BC) was the 'father of botany'. The collection and classification of species has been a major activity amongst naturalists over the years, particularly those who lived during and after the Scientific Revolution.

It is an interesting coincidence that the person who essentially established the foundations of the study of the world of life lived and worked at the time of Newton. The times were obviously ripe for innovation and new fields of science. John Ray (1627–1705) became the most prominent naturalist of the seventeenth century. He was the son of a village blacksmith, and managed to make his way to Trinity College, Cambridge. His talents were clear, and he was made a Fellow of Trinity in 1649. The Fellows were free to study whatever they liked, and for Ray it was botany. As there was at that time no classification scheme for plants, he set out to make one. Fortunately for science, he eventually teamed up with a colleague, Francis Willughby (1635–1672), who shared many of his interests. They made several field trips, not only in England but also in much of continental Europe, Ray concentrating on plants and Willughby on animals. The European trip is said to have been for them what the voyage of the *Beagle* was for Darwin. They returned with a vivid knowledge of the living world and a huge number of specimens and notes. Willughby died in 1672, but Ray persevered with their planned publications. *Ornithology* and the *History of Fishes* were published under Willughby's name in 1677 and 1686, and the three immense volumes of Ray's *History of Plants* were published in 1686, 1688, and 1704. His *History of Insects* was published posthumously in 1710. These works paved the way for the more famous Carl Linnaeus.

Ray was one of the first to recognize fossils as the remains of once-living plants and animals, and he pondered over the idea that entire species may have disappeared from the planet. He also wondered about the implications of the presence of fossils of fishes found in high mountains, and the long timescales likely required for the mountains to have risen so high.

Meanwhile, the microscope, developed in the early seventeenth century, provided a range of amazing discoveries in biology, from the structure of a fly's eye to the cells of living matter and the first glimpses of microbes. The promise of this new technology was highlighted by the wonderful illustrations in Robert Hooke's book *Micrographia*, published in 1665. Over time microscopes became more powerful and reliable, and they eventually played an essential part in elucidating the very processes of life, as we shall see later.

Carl Linnaeus (1707–1778) is famous for having given us the classification scheme for plants and animals. He was born in southern Sweden, and studied medicine at the Universities of Lund and Uppsala. Since childhood he had an interest in flowering plants, which he studied alongside his courses in medicine. He went on several botanical expeditions before receiving his medical degree in 1735, and was eventually appointed to the chair of botany at Uppsala in 1742. He was an inveterate classifier. While he was still a student he published his ideas on taxonomy in *Systema Naturae* in 1735. That work was revised, updated and published in ten editions, the last in 1758. In his system the hierarchy of life (top-down) is kingdom, phylum, class, order, family, genus and species. Linnaeus introduced a binomial nomenclature, giving every living organism a two-word name (genus and species). From the results of his own field trips and those of others, Linnaeus classified thousands of plants and animals. His system is still used today.

Significantly, Linnaeus was bold enough to include mankind (*Homo sapiens*) in his system, under Animalia. He commented that he could see no scientific reason to separate mankind from his classification system, which included the apes.

A topic related to the evolution of life was the age of the Earth. In 1620 the Irish Archbishop James Ussher (1581–1656) calculated from the Bible that the year of the Creation was 4004 BC. By the eighteenth century this was already coming into question from a variety of directions. Knowledge coming from China indicated that the first Emperors dated from about 3000 BC, and Chinese history must have gone back considerably before that. Ray noted the likely conflict between the ages of the mountains and the biblical time scale. Linnaeus also had doubts, from the existence of fossils far from any present-day seas. Newton himself had said that a globe of red-hot iron the size of the Earth would have taken over 50,000 years to cool. At least one person tried to do the experiment using hot balls of iron: Georges Louis Leclerc (Comte de Buffon, 1707–1788), widely known for his monumental 44-volume *Histoire Naturelle*, estimated that the Earth must be at least 75,000 years old. Another Frenchman, Jean Fourier (1768–1830), famous for the mathematical Fourier transform, used equations of heat flow to estimate an age of 100 million years. And the fossil evidence seemed to be pointing to a very long time span.

Georges Cuvier (1769–1832) became probably the most influential biologist in the world in the early 1800s. He was on the staff of the Museum of Natural History in Paris for much of his life. He worked on comparative anatomy, and published a five-volume work on that subject. He highlighted the differences between the anatomy of meat-eating and plant-eating animals, and was then able to distinguish fossils on that basis. He believed that species, once created, remain fixed in the same form until they become extinct. He laid the basis for the science of palaeontology. It became possible to place the strata in which fossils were found in chronological order. He believed that the Earth had experienced a series of catastrophes which led to extinctions, and, like others, thought that the history of life went back a very long way in time.

A protégé of Cuvier, Jean Baptiste Lamarck (1744–1829) also worked at the Museum of Natural History, and became famous for his idea that characteristics can be acquired by individuals during their lifetime and then passed on to the next generation. The classic example is the giraffe, which (according to Lamarck), in stretching for the topmost leaves on a tree, actually lengthens its neck over its lifetime; its offspring are then born with longer necks as a result ('Lamarckian inheritance'). Unlike Cuvier, Lamarck thought that species never become extinct—they just change into another form. These ideas were taken up by Étienne Geoffroy Saint-Hilaire (1772–1844), who went a step further. He proposed that the environment may play a direct role in evolution: if the Lamarckian modifications lead to injurious effects, those individuals will die off and be replaced by others better suited to their environment. To our modern ears, this almost starts to sound a bit like Darwinism.

In the late eighteenth century there was a growing awareness of the geological processes that have shaped the Earth and determined the environments to which species had to adapt. James Hutton (1726–1797) is known for the principle of uniformitarianism—that if enough time were available, all the features of the world around us could be explained in terms of the same geological processes we are familiar with today—continuous erosion and uplifting, with occasional earthquakes and volcanic activity. Nothing as extreme as the violent convulsions envisaged by the alternative view known as catastrophism. But uniformitarianism did require the age of the Earth to be immensely greater than previously envisaged. Charles Lyell (1797–1875), initially a lawyer who became fascinated by geology, made a historic geological expedition through Europe in 1828, and came back sharing Hutton's view. And he was impressed by the thought of how life would have to adapt over the long term to the ever-changing environment caused by the geological processes. His book *Principles of Geology* was very popular and influential. In the second edition he wrote that many species that once lived on Earth apparently became extinct and replaced by other species, and that the reason they became extinct may have been competition for resources.

The scene was now perfectly set for Charles Darwin (1809–1882). He was born near Shrewsbury into the family of a medical doctor, with three sisters and a brother. His early years were happy, until the death of his mother when he was 8 years old. He was sent to a nearby boarding school, where he developed a strong interest in natural history. At the age of 16 he was sent to medical school in Edinburgh, but he apparently hated the sight of blood and changed to courses in natural history. Two years later he was moved to Christ's College in Cambridge, this time to work on a degree in preparation to be a country clergyman. Again he switched to courses in natural history, and was considered an outstanding student. In spite of his neglect of the conventional curriculum, he managed to obtain a respectable degree and graduated in 1831. And then, out of the blue (through Cambridge connections), he received an invitation to accompany Captain Robert FitzRoy on the HMS *Beagle* for a 5-year voyage around the world and to study the natural history and geology of South America in particular.

They set sail in 1831, when Darwin was just 22. It was an epic voyage, with a great many adventures and experiences to feed his fertile mind. The *Beagle* was to survey the coast of South America, but Darwin was free to explore, and spent most of the time on land expeditions. He saw tropical forests in Brazil, rode with gauchos into the interior of Argentina, experienced an earthquake in Chile, explored the wildlife of the Galapagos islands, visited Tahiti, saw the marsupials of Australia, explored the atolls of the Cocos (Keeling) islands, and visited Mauritius. He developed ideas about uplift following an earthquake, the variety of related species on the different islands of the Galapagos, and the formation of coral atolls. His specimens and letters were sent back to England at intervals, so that when he returned in 1836 he was already well known in scientific circles.

Darwin described the period between 1836 and 1842 (when he and his new wife Emma moved from London to their new home in Kent) as his most creative period intellectually. During that time he interacted with many of the scientific luminaries of the land including Lyell himself, gave lectures, and received acclaim as a writer. In 1839 his *Voyage of the Beagle* was published, he was elected a Fellow of the Royal Society, and he was married. His father made financial arrangements so that Darwin could continue his work as a self-funded 'gentleman scientist' for the rest of his life. Darwin and his wife settled down in their home in Kent, and had ten children. Darwin was free to study, write, and correspond with other naturalists.

By the time the *Beagle* returned to England, Darwin was convinced that evolution was a fact. But what caused it? Darwin started his first notebook on *The Transformation of Species* in 1837. A major influence on Darwin was the book *Essay on the Principles of Population* by Thomas Malthus (1766–1834). Malthus showed that populations have the power to grow exponentially unless they are checked by predators, diseases and shortages of food. Darwin realized that this could be the key: competition within and between species could result in only the fittest (best-adapted) being able to reproduce and survive. Those less fit would die and become extinct, to be replaced by those better adapted to their circumstances ('survival of the fittest'). This became the theory of evolution by natural selection; it was already outlined by Darwin before he moved to Kent. In 1844 he wrote a manuscript giving his ideas, accompanied by a letter to his wife requesting that it be published after his death. He didn't rush to publication because he was concerned about the possible reaction of the Church and the public, and he also worried about upsetting Emma, a devout Christian. Instead, he largely turned his attention to a new project: barnacles. He completed a three-volume work on the topic in 1854, for which he was awarded the Royal Medal from the Royal Society. In that same year he started to assemble his notes and manuscripts in preparation for a massive work on the evolution of species.

In 1858, Darwin was shocked to receive a paper entitled *On the Tendency of Varieties to Depart Indefinitely from the Original Type* from Alfred Russel Wallace (1823–1913). Wallace had a great interest in natural history, and made a living of sorts by collecting and selling specimens in Brazil and the Far East to museums and wealthy collectors in England. He had met Darwin once, and the two began a correspondence; Darwin even became one of Wallace's customers. Wallace developed his own initial ideas on evolution, and published them in a paper in 1855. Darwin's friends became concerned that Darwin might be pre-empted, and urged him to publish. Like Darwin, Wallace was motivated by the work of Malthus, and in 1858 he had the same insight that Darwin had in 1837. When Darwin received Wallace's paper in 1858, asking for his opinion, he did the right thing by passing the correspondence to Lyell, offering to send Wallace's paper to a journal. Lyell had a different idea—that Darwin's 1844 outline be added to Wallace's paper and offered to the Linnean Society as a joint publication. The paper was read to the Society and published without causing a stir. Wallace was delighted by this outcome, and grateful to Darwin.

Darwin then made a huge effort to complete his *magnum opus*, *On the Origin of Species by Means of Natural Selection, or the Preservation of Favoured Races in the Struggle for Life*, which was published to great acclaim in 1859. It was widely accepted because it was so well argued and so well documented by examples of all kinds. It changed the field of natural history forever.

There were still, however, two lingering questions about Darwin's theory that were debated throughout the later years of the 1800s. One concerned the very long timescale required for evolution. The physicists, in particular William Thomson (Lord Kelvin), argued that, according to the physics of the time, the Sun would not be able to continue shining for anything close to what was required. This matter was resolved in the early decades of the twentieth century, when it was realized that sub-atomic processes could keep the Sun going for many billions of years. The other problem was that the mechanism for heredity, required for evolution by natural selection, remained unknown.

As it turned out, the answer to this second issue was already being worked out by a Moravian friar, Gregor Mendel (1822–1884). Mendel was born into a poor farming family, which stretched its resources to educate him. He graduated through high school (gymnasium), and studied philosophy and physics for 2 years at the University of Olmuetz. By then he had exhausted his financial resources, and became a friar at the monastery of St. Thomas in Bruenn to extend his education. Following completion of his theological studies he was sent to the University of Vienna for 2 years, where his courses included physics, statistics, chemistry and plant physiology. He returned to the monastery in 1853 as a teacher. He had ample time for research, and in the period 1856–1868 (after which he was elevated to abbot) he did ground-breaking work on heredity in peas.

Mendel had a large plot of land in the monastery's experimental garden, and worked with thousands of pea plants. His work required great care, discipline, accurate records and good statistics. Each of his experimental plants was pollinated by hand. His work clearly revealed the role of certain factors in determining a plant's characteristics (its 'phenotype'). Today we call these factors genes. They come in pairs; one is dominant in being expressed in the phenotype, and the other is recessive. The results of his studies showed that one in four of the pea plants had purebred recessive genes, one in four were purebred dominant, and two in four were hybrid. He was able to make generalizations which became known as Mendel's Laws of Inheritance. He showed that inheritance works not by blending characteristics from two parents, but by taking individual characteristics from each of them. He was fortunate in having chosen peas which are so simple genetically; inheritance in most organisms is more complicated.

Mendel presented his results to the Natural Science Society in Bruenn in 1865, published them in a paper (in German) in 1866 in *Verhandlung des naturforschenden Vereins Bruenn*, and sent copies to leading biologists around Europe, but aside from some local newspaper reports it had little impact. It is a great pity that neither Darwin nor his associates became aware of Mendel's work[11]—it would have provided the mechanism for his theory, and would have advanced the science of genetics by over 30 years. Nevertheless, his laws (and his paper) were rediscovered by four others at the turn of the century. The Dutch botanist Hugo de Vries (1848–1935) carried out very similar research, and in 1899 was about to publish the results of his own work when he came across Mendel's paper. The German Carl Correns (1864–1933), the Austrian Erich von Tschermak (1871–1962), and the American William Spillman (1863–1931) suffered the same fate, but all four of them gave credit to Mendel as the real discoverer of the laws named after him.

However, the underlying chemical basis was not yet known. It had long been established from studies using the microscope that all living things are made up of cells, and that all cells are produced from other cells by division. The cell was known to be composed of an outer wall enclosing a viscous liquid of some kind and a central concentration called the nucleus. Researchers had observed the penetration of the sperm into the egg, with the two nuclei fusing into one. In 1879 Walther Flemming (1843–1915), using coloured dyes to highlight the internal features of cells, found that the nucleus contains distinct thread-like structures, which he called chromosomes. He saw that, when a cell divided, the chromosomes were duplicated and shared between two daughter cells. August Weismann (1834–1914) noted the difference between the process of cell division for growth and that for producing eggs or sperm.

The American Thomas Morgan (1866–1945) laid the foundation for both the science of genetics and evolution by natural selection. He worked with the common fruit fly *Drosophila*, which had the dual advantages of a rapid breeding cycle and only four chromosomes. He found several stable heritable mutations in *Drosophila* which provided vital clues to heredity and evolution, and he and his colleagues wrote the very influential book *The Mechanism of Mendelian Heredity* in 1915. Darwin's theory now had a solid genetic foundation, and the next step was to determine the nature of the chromosomes themselves.

---

[11] In fact, Darwin was one of the biologists to whom Mendel had sent his paper; that copy (with its double pages still uncut, indicating that it had not been read) was found years later in Darwin's vast library of books and papers.

It was known that the chemical composition of the nucleus was different from that of proteins. In 1869 the Swiss biochemist Friedrich Miescher (1844–1895) was able to extract what he called nuclein (now known to be deoxyribonucleic acid, or DNA) from the nucleus, and showed that it was an acid containing phosphorus, and therefore unlike the known groups of biological molecules such as carbohydrates and proteins. In 1885, Oskar Hertwig (1849–1922), aware of the studies of Flemming and Weismann, went so far as to declare that "Nuclein is the substance which is responsible... for the transmission of hereditary characteristics". Many disagreed. Chromosomes also contain proteins, and it was thought that only proteins could carry the code of life.

Phoebus Levene (1863–1940) studied the structure and function of nucleic acids, and showed that the components of DNA are linked in the order phosphate-sugar-base. He held that DNA is a relatively small molecule, and that it is organized in such a way that it could not carry genetic information. His ideas were widely accepted, and supported the view that the important information in the cell is contained in the structure of the complex proteins, with DNA just playing a supporting role. But in a major development in 1944, Oswald Avery (1877–1955) and colleagues solved a puzzle that involved a genetic transformation phenomenon between two types of bacteria that cause pneumonia; they were able to prove that DNA, not proteins, was responsible. So perhaps the genetic material is DNA after all?

The paper by Avery and colleagues stimulated Erwin Chargaff (1905–2002) to study DNA in individuals and species. He found that the composition of DNA is identical for all tissues within a given species, but that it differs from one species to another—just what would be expected if it is the biological signature of the species. DNA contains 'base molecules' called A, C, G and T, and Chargaff also showed that the amount of the base A in a sample of DNA is the same as that of T, and the amount of G is the same as that of C. These relations became known as the Chargaff rules, and they were crucial in elucidating the structure of DNA.

In the late 1940s and early 1950s there were three groups working on the structure of DNA: Linus Pauling (1901–1994) at Caltech, Maurice Wilkins (1916–2004) and Rosalind Franklin (1920–1958) at King's College London, and Francis Crick (1916–2004) and James Watson (1928–) at Cambridge. Pauling was one of the most famous biochemists of his time, and was now turning his attention to DNA. He had discovered helical forms in proteins, and so would certainly be attuned to hints of helicity in DNA. In early 1953 Pauling announced that he had found the structure of DNA—a triple helix. When Crick and Watson saw a preprint of the paper, they knew immediately that Pauling's result was wrong, as it was inconsistent with X-ray crystallography data being obtained by Franklin, but also that they would have to rush with their work if they were to be the first to find and publish the correct structure.

Crick was a physicist who had turned to biology, and Watson was an ambitious young American PhD in molecular biology who had DNA in his sights. They shared an office in Cambridge, and started an unofficial collaboration to work out the structure of DNA using a model-building approach. It was clear to them that the structure had to allow for an exact copying process so that an identical copy could pass into the daughter cells. They were also very aware of the rules found by Chargaff, who visited Cambridge in 1952. When Watson showed a copy of Pauling's paper to Wilkins in early 1953, Wilkins responded by showing Watson a print of one of Franklin's best X-ray images, without her knowledge. This breach of etiquette was crucial in leading Crick and Watson to the correct solution—a double helix. They published their paper, *A Structure for Deoxyribose Nucleic Acid*, to great acclaim in April 1953. It was a monumental discovery.

Franklin was actually close to publishing her own version of the double helix when the news of Crick and Watson's result came down from Cambridge. Franklin and Wilkins did receive some credit for their work: their own separate papers were published together with that of Crick and Watson in the same issue of *Nature*.

The simple and elegant double helix structure immediately suggests its function. It is ideal for all three functions of DNA—telling cells what they have to be and do, providing the basis for cell reproduction, and passing the genetic code on from generation to generation. Inheritance based on DNA readily provides the mechanism for evolution by natural selection. It is astonishing that all the complexities of life are based on this simple double helix, which is common to all life forms.

But how did this actually work? George Gamow suggested that the genetic code is written in triplets along the length of the DNA strands. A single triplet, containing three positions each of which can be occupied by one of four possible bases, allows for 64 possible combinations. That is more than enough to account for the 20 amino acids that comprise proteins. Then how is the information encoded in the DNA used to make the amino acids? Crick proposed that some unknown 'adapter' was involved in conveying the information from the DNA. Marshall Nirenberg (1927–2010) and a colleague were the first to work out the genetic code corresponding to a specific amino acid—a major and famous breakthrough—and Robert Holley (1922–1993) discovered transfer RNA, the 'adapter' that Crick had proposed. These and other scientists succeeded in breaking the entire genetic code for protein synthesis. We could now read the messages encoded in the DNA!

Now that the structure and function of DNA was known, a long-term objective was to map out the entire human genome—the sequence of the positions of all three billion base pairs along the strands of our DNA. The first step was to cut out and make many copies of small stretches of DNA; this would be done by enlisting the help of special enzymes for cutting, and bacteria to make many copies. Then it was necessary to find the order of the base pairs in these sections of DNA, and piece all the information together.

The crucial enzymes that allowed the DNA to be cut and manipulated were discovered by Werner Arber (1929–), Daniel Nathans (1928–1999) and Hamilton Smith (1931–). These 'restriction enzymes' can recognize a specific sequence or site on the DNA and cut the genome into short strands. The ends of these strands can then be tagged with a radioactive label so that they can be identified.

In a major breakthrough in 1977, Frederick Sanger (1918–2013) and his colleagues developed a rapid sequencing technique which allowed long stretches of DNA to be sequenced. This became known as the 'Sanger method'. (Sanger had previously shown in the 1950s that every protein has a unique genetic sequence—crucial for solving how DNA codes for proteins).

Following several other technical steps along the way, the very ambitious international Human Genome Project to determine the DNA sequence of the entire human genome was launched in 1990. It was completed in 2003 under the leadership of Francis Collins (1950–), Director of the NIH (U.S. National Institutes of Health), two years ahead of schedule. It had cost three billion dollars, and involved massive sequencing and computing power at 20 institutes in six countries. It remains the world's largest collaborative project in biology. (A smaller and cheaper private competing project, aided by data from the publicly funded project, was completed at the same time by Craig Venter (1946–) and his firm Celera Genomics). The sequencing of the human genome was a huge success to cap off centuries of searching for the secrets of life.

Since then the genomes of many other species have been sequenced. A Japanese flower, *Paris japonica*, has the longest known DNA with 149 billion base pairs, wheat has 15 billion, the human genome has three billion, and the smallest non-viral genome is that of a bacterium with 160 thousand base pairs, so ours is by no means exceptional in terms of length.

While the story of DNA has been truly spectacular and fundamental, it is only one part of molecular biology. The cell is vastly more complex. A few of the more important developments in molecular biology over the past several decades are summarized in the following paragraphs, to give an idea of the complexity and the progress.

In the late 1930s Conrad Waddington (1905–1975) considered how genes could produce developmental phenomena, and wrote a book on 'epigenetics' (the prefix 'epi' signifies 'over' or 'above'—epigenetics is the regulatory system superimposed over the genome itself). He coined the term 'epigenotype', which refers to all the complex entities and interactions that operate between the genome and the phenotype. He also introduced the very influential concept of the 'epigenetic landscape', which gives a metaphorical image of how gene regulation modulates development such as cell differentiation.

It was clear that a knowledge of the mechanisms of gene expression would be of critical importance. In 1961 Francois Jacob (1905–1975) and Jacques Monod (1910–1976) studied the bacterium *E. coli*, and found that there are specific proteins that can suppress the transcription of the relevant genes—a feedback loop was found. This crucial discovery opened up the study of the principles of gene regulation—how genes are expressed. Regulatory phenomena are now found at every level of gene expression.

In 1967 Mark Ptashne (1940–) isolated a protein from a virus that infects bacteria using the bacterium's DNA to reproduce itself. The protein works by attaching to a specific site on the bacterial DNA. With others, Ptashne found that proteins regulating genes work like keys that fit specific locks. After this binding occurs, the protein key interacts with other binding proteins to turn the gene on or off. He and his colleagues later found that the same mechanism works in other organisms, such as yeast, fruit flies, plants and humans—it explains gene activation throughout nature.

Sydney Brenner (1927–) made several important contributions to molecular biology and animal development. His insights in the early 1960s led to the central paradigm of molecular biology—that information flows from the DNA to proteins and never the other way around. He introduced the concept of the 'messenger RNA'. He has been quoted as saying that "the questions facing pre-genomic and post-genomic biology are still the same—bridging the gap between the genotype and the phenotype".

An entirely new field of molecular biology was suddenly opened up in 1998 by Andrew Fire (1959–), Craig Mello (1960–) and colleagues, when they published a paper in *Nature* showing that there are large numbers of tiny snippets of double-stranded RNA that can destroy the messenger RNA before it can produce a protein, effectively shutting the gene down. The process is known as RNA interference, or gene silencing by double-stranded RNA. This important work has revolutionized knowledge about processes and regulation in molecular biology.

In 1958 John Gurdon (1933–) successfully cloned a frog using intact nuclei from the somatic (body) cells of a tadpole. From this he concluded that differentiated somatic cells have the potential to revert to pluripotency. This work became well known, and his techniques are still used today. At that time he could not prove that the transplanted nuclei had come from a fully differentiated cell, but that was finally confirmed in 1975. It produced a major change in how geneticists thought about cell differentiation. Much later, in an astonishing breakthrough in 2006, Shinya Yamanaka (1962–) and colleagues were actually able to make pluripotent stem cells from normal somatic tissue in adult mice using a novel technique involving what are essentially epigenetic regulators. (Pluripotent cells can give rise to any and all of the different types of cells that make up the body; stem cells are pluripotent.) This opened up a potentially very important way of *reversing* the effects of ageing.

And indeed, in a subsequent breakthrough in 2016, Juan Carlos Izpisua Belmonte (1960–) and colleagues used the regulating 'Yamanaka factors' to partially 'rewind the epigenetic clock' in the cells of mice that age prematurely; they actually increased the lifespans of these mice by 30%, clearly demonstrating that ageing is not an irreversible process. They had been inspired by their studies of the regeneration of lost tails or limbs by some lizards and fish. In a 2018 review paper[12] they suggested that ageing, which is an inevitable decline due to the relentless tug of entropy and is the major risk factor for most of our diseases, might *in principle* be slowed by periodic in vivo epigenetic 'reprogramming' of the 200 cell types in our bodies. Their emphasis is on increasing the healthy years of life, but such work obviously also has implications for overall human longevity. Meanwhile, other researchers are taking various different approaches to the problem of ageing, which now begins to seem less intractable than it did in the past.

Molecular biology is endlessly complex and fascinating, and a wealth of knowledge has been created. But there are also several other fields of biology that exploded onto the scene in the twentieth century.

Neuroscience is one of them. Camillo Golgi (1843–1926) and Santiago Ramón y Cajal (1852–1934) did early ground-breaking work on the structure of the nervous system. Golgi discovered a method of staining nervous tissue which made it possible to see individual neurons and their unique and remarkably complex structures in the brain for the first time. From his work he concluded that the nervous system is a single network. Ramón y Cajal is

[12] Beyret et al. (2018) Elixir of Life: Thwarting Aging with Regenerative Reprogramming.

often referred to as 'the father of modern neuroscience'. He improved and used Golgi's staining method, and his detailed drawings of neurons became famous. He discovered that neurons exist as separate entities, rather than as nodes in a continuous network as Golgi had suggested. Their relationship to each other is contiguous, not continuous. This was called the neuron theory, and it became the foundation of modern neuroscience.

In the late 1930s Alan Hodgkin (1914–1998) and Andrew Huxley (1917–2012) developed methods that enabled them to record the currents in live axons using the crude equipment available at that time. Following the pioneering work of John Young (1907–1997), they used the giant axon of the Atlantic squid, which has the largest neurons known, in order to provide ionic currents that could be measured. In 1939 they were able to report the first detection of 'action potentials', the electrical signals which enable the activity of an organism to be coordinated by a central nervous system. After the war they continued this work, and were able to show just how action potentials are transmitted. Their hypothesis of the existence of 'ion channels' in cell membranes was confirmed decades later.

Karl Lashley (1890–1958) was one of the world's foremost brain researchers, conducting experiments and studies on the brain from the 1920s to the early 1950s. He measured the behaviour of trained rats before and after specific induced brain damage, trying to find a single biological centre of memory. In the end he concluded that memories are not localized in one part of the brain, but rather spread throughout the cortex. This influential work led to an anti-localization bias amongst scientists that lasted for decades. In fact, we now know that most brain tissue is highly specialized, although a typical cognitive act does indeed activate many regions of the brain.

Knowing a neural circuit does not necessarily tell us how it works. Eve Marder (1948–) studied a very simple circuit of just 30 neurons involved in the control of the digestive system of the crayfish. She discovered that the connections and behaviour of the cells are modulated by a 'soup' of dozens of different chemicals, changing the functionality of the circuit. She showed how the brain can change during development while remaining structurally stable. It has been said that such complexity coming from an apparently very simple circuit can be considered a metaphor for all of biology.

In an amazing recent study, David Glanzman and colleagues at UCLA succeeded in 'transplanting' memory from one animal to another. They trained some snails with traumatic shocks, so they learned to withdraw into their shells for 50 seconds instead of the usual one second. They then injected RNA from these snails into other, untrained snails, which subsequently withdrew for similarly long periods on sensing a tap. Aside from being a dramatic new development, this is consistent with the idea that memory involves RNA-induced epigenetic changes in the neurons.

Mathematics was introduced into biology early last century. The three major figures to develop the mathematical theory of population genetics were Ronald Fisher (1890–1962), John Haldane (1892–1964), and Sewall Wright (1889–1988). They used mathematics to combine Mendelian genetics and natural selection, which became known as the neo-Darwinian modern evolutionary synthesis. The mathematical tools produced by Fisher in particular created the foundations for modern statistical science, not just in biology but also in psychology and several other fields. All three of them made other important contributions to evolutionary science; Haldane considered the possible origin of life from inorganic molecules, and introduced the concept of the 'primordial soup', which later became known as the Oparin-Haldane hypothesis.

William Hamilton (1936–2000) was one of the most important evolutionary theorists of the last century. He dealt with a problem that had worried both Fisher and Haldane: how an organism could increase the fitness of its genes by helping its relatives at its own expense—he had produced a rigorous genetic basis for the existence of altruism. His two 1964 papers entitled *The Genetic Evolution of Social Behavior* and his 1970 paper *Selfish and Spiteful Behaviour in an Evolutionary Model* are considered foundational. His was one of the forerunners of sociobiology, and contributed much to evolutionary biology and behavioural ecology. Edward Wilson (1929–) is considered 'the father of sociobiology', and in 1975 he published the book *Sociobiology: The New Synthesis*, applying his theories of insect behaviour to vertebrates and humans. His inclusion of humans created a storm of protest at the time (some protesting that humans are not animals). Richard Dawkins (1941–) popularised these ideas and gave a gene-centric view in his 1976 book *The Selfish gene.*

These are just a few of the developments in some of the fields of biology over the past several decades. Biology continues to grow on all fronts, and it is certainly now one of the biggest areas of science.

2.12 Our Evolving Perspectives

Looking back over this chapter, we can clearly see the broad contours of the history that led to modern science.

Our ancestors lived as simple hunter-gatherers for 99.9% of the time since they speciated away from the chimpanzees some 7 million years ago. Change was glacially slow. Millions of years elapsed before they developed stone tools, and another million years passed until they had the use of fire. Tens of thousands of years ago they began to develop a host of innovations, ranging from improved weapons, jewellery, cave paintings and symbols to burial rites. After millions of years of evolution they were finally becoming similar to modern humans.

Ten thousand years ago they developed agriculture, and their world changed. They were able to live in settlements, populations grew, and the first major civilizations arose about five thousand years ago. There were developments based on elementary science—astronomy, engineering and other fields—but none of these civilizations produced natural philosophy, the rational study of the natural world and the basis of modern science. Why? The main reason was undoubtedly the existence of well-established religions which claimed to explain the world in terms of gods and myths, apparently leaving nothing more to be done. And their imposing priesthoods, deeply embedded in the state hierarchy, would have deterred any attempt at free thinking.

But in the sixth century BC the first of the two fundamental steps in the Rise of Science took place—the 'Greek Miracle'. The world was to be explained by rational thought in terms of causes that were part of nature itself, rather than by religions, gods and myths. This was a revolutionary development. Why Greece? It was comprised of decentralized city-states, and the citizens were relatively free. There was certainly religion in Greece, but it was fragmented, with no overall priestly caste to impose dogma. Debate was customary, and novel ideas could flourish. The philosopher who had the crucial insight was Thales of Miletus, who was educated, worldly-wise and a polymath. He is considered to be the father of natural philosophy.

This philosophical tradition in Greece lasted for a thousand years, peaking around 300–500 BC, the time of Socrates, Plato and Aristotle. It was Aristotle who made by far the greatest contributions to natural philosophy, and his works dominated the scientific landscape for almost two thousand years. But this Greek philosophical tradition gradually faded away, for no obvious reason. Perhaps it was felt that everything that could be said had been said—that the wisdom of the towering figures of the classical time could never be surpassed. The later years overlapped with the Roman Empire, which had no time for such frivolous ponderings, and the new cult of Christianity was antagonistic to 'pagan' studies. By 500 AD the Great Library of Alexandria had been destroyed and Rome had fallen.

But in the seventh century there was an important new development: the rise of Islam. There was considerable interest in the Greek classics, and a major translation effort into Arabic took place. Following in the Greek tradition,

Islamic science, especially in astronomy, mathematics and medicine, flourished during the Islamic Golden Period, which peaked around 1000 AD but then declined as attitudes hardened against 'foreign' and non-Islamic studies.

Meanwhile, in Western Europe the Greek classics were being translated into Latin, some from the original Greek and many others from the Arabic copies that became available. Monasteries played a significant role in this activity. In 1100–1200 several universities were established in Western Europe, and the curricula concentrated on the Greek classics. Over the next few centuries a number of medieval and Renaissance European thinkers at these universities went beyond the level of the original Greek natural philosophers, but by 1600 the prevailing worldview still remained that of Aristotle and Ptolemy.

In the sixteenth and seventeenth centuries the second fundamental step in the Rise of Science took place—the Scientific Revolution. Copernicus changed the centre of the cosmos from the Earth to the Sun, Kepler and Galileo gave convincing evidence supporting this heliocentric concept, and Newton proved the unity of Earth and cosmos and established the physical laws that govern everything. Events could be predicted, and with astonishing accuracy. It was a monumental step. Aristotle and Ptolemy were relegated to the history shelves. The world works in accordance with fundamental physical laws which were proven again and again. It was a stunning and complete revision of our view of the world. Modern science was born.

In the 1700s and 1800s this worldview became even more firmly established. The term natural philosophy was gradually replaced by the word 'science' to refer to studies of the natural and physical world. Faraday's experiments and Maxwell's equations clarified and unified electricity and magnetism, Darwin presented his compelling theory of evolution by natural selection, and physics seemed to be just about complete—there didn't seem to be much more to do.

But then, in the early 1900s, more revolutions took place. Einstein's theory of relativity replaced Newton's concept of absolute space and time, and quantum mechanics emerged to form the basis of our knowledge of the atomic and subatomic world. The genetic basis of life became known. Our universe was found to be vastly bigger than expected, and its expansion was discovered. We learned to fly, we invented radio, television, atomic power and the Internet, and we went to the Moon. The increase in our scientific knowledge and technological capabilities over just the last few thousandths of one percent of our existence has been spectacular.

The fact that natural philosophy arose only once—in Greece—highlights its uniqueness and importance: it was a precious treasure for all humanity. It was carried in a fragile line through history, from the Greeks via the Islamic and medieval periods to the Scientific Revolution to today. And now it is accessible to the entire world.

It is interesting to note that there was no overall plan for the development of science. Individuals simply added in their own way to the knowledge of their time, and the net result over centuries is the great body of scientific knowledge that we have today.

So, over the entire history of science there were just two major revolutions: the advent of Greek natural philosophy in the sixth century BC introducing the concept of natural causes, and the Scientific Revolution in the seventeenth century introducing quantitative predictions and testable laws of nature.

What would have happened if these two fundamental steps in the Rise of Science had not occurred? One only has to look at the development of the rest of the world over most of that period. In a few parts of the world some hunter-gatherer societies still exist. In many other areas people eked out a meagre existence in the shadows of the giant structures of long-past civilizations until recently. China has had a long record of innovations, but until recently the bulk of the population lived as poor peasants and villagers. Japan only emerged from its feudal society in the late 1800s.

Would the Industrial Revolution in Great Britain still have taken place? There are thought to have been various causative factors; certainly a major one was the stimulation provided by the profoundly new worldview of the Scientific Revolution. The world could be explained, predicted and manipulated for the benefit of mankind. This was an inspiration for The Age of Enlightenment, which was rational, optimistic and progress-oriented, and would have been a huge spur for the energy and enterprise of the Industrial Revolution. While the innovations in the early stages of the Industrial Revolution did not directly involve advanced scientific knowledge (the first steam engine was actually produced by the ancient Greeks), developments after 1800 (such as the telegraph) did require more than elementary scientific knowledge, and advanced science rapidly came to play a major role in later developments.

The remarkable revolution in public health throughout much of Europe in the mid 1800s also did not initially depend directly on advanced science. Long gone were the days when a disease or epidemic was merely attributed to superstition or religion, and rational and common sense approaches were taken to understand causes and find solutions. The public health initiatives of the mid 1800s were largely responsible for the significant increases in health and longevity over that period. But in the late 1800s a truly scientific revolution took place in medicine as the actual causes of disease were identified in the laboratory, and 'miracle cures' became possible.

In general, the application of advanced scientific knowledge to technology and medicine accelerated rapidly through the 1800s, creating a whirlwind of innovations that accelerated towards the end of that century and into the next. The result has been unprecedented exponential growth in both science and technology over just the last two or three lifetimes—a tiny fraction of human history.

Today the benefits of the scientific revolutions are obvious. Without them the world would look much as it did hundreds or even thousands of years ago.

And it is not only technology and living standards that are so much greater today. Our knowledge of the world is vastly greater than it was just a few hundred years ago. Today we know about the basis and unity of all life and how it evolved, we know about the atom and its constituents and have unleashed the potential of nuclear power, and we understand the universe, its evolution and contents. Our knowledge today is vast, and we have really only started to make use of it. The future of science and technology is bright indeed.

3

Roads to Knowledge

The broad historical overview given in the last chapter of many of the scientists and their achievements provides an opportunity to take an orthogonal approach and look at the variety of ways in which the individual scientists actually did their work and achieved their great results. Were they alone or in groups? Was it pure inspiration and creativity, or dogged slogging to collect data and forge hypotheses? Did they know what to look for, or did they stumble onto discoveries? Were they just in the right place at the right time? The roads to knowledge are not always as neat and tidy as the classical textbook scientific method would suggest.

So what is science really like? In this chapter it will be shown, using individual stories and anecdotes, that science can be as complex as any human activity, and that any of a variety of characteristics, factors and methods can be involved in the scientific process. Some of them, in no particular order (and sometimes overlapping, sometimes contradictory), are self-motivation, intelligence, passion, pragmatism, freedom, connections, mistakes, experimentation, curiosity, teamwork, education, discovery, observation, deduction, persistence, socializing, judgment, lateral thinking, cross-pollination, creativity, solitude, imagination, inventiveness, collecting, synthesis, tenacity, intuition, opportunity, insight, falsification, technology, instinct, scepticism, disagreements, open mind, honesty, timing, serendipity, determination, verification, exploration, interpretation, communication, inspiration, induction, experience, theory, speculation, hypotheses, reflection—and there are many others.

P. Shaver, *The Rise of Science*, https://doi.org/10.1007/978-3-319-91812-9_3

Curiosity, intelligence, freedom, education, self-motivation and determination are clearly essential. Luck and serendipity can sometimes play a role—stumbling onto a discovery, or finding a vital clue that leads to a major development. Mistakes can happen, and can sometimes end up being beneficial. Many of the great scientific results come from dogged and time-consuming experimentation, observation and sample collecting. Connections can be very important—others who put the scientist on the right track or in the right role. Even ignorance can sometimes be a factor; the scientist blithely carries on with a course of investigation not knowing that others would say it can't be done, and stumbles across an important result that surprises everyone. Solitude is sometimes essential for peaceful reflection and inspiration; in other cases teamwork is helpful or even essential.

Lateral thinking (or 'thinking outside the box') often leads to new results. Cross-pollination between different disciplines can bring in fresh ideas and technologies, so it helps to know people in a variety of fields; the results can sometimes be revolutionary. Imagination can lead to entirely new avenues of science. Arguments and disagreements are natural and important ingredients of the scientific process: different views are debated, and the result is increased rigour in our knowledge of the subject under investigation. By far the greatest critics of any scientist are other scientists.

Most of the science we do today depends on technology—the tools we use to make our discoveries and observations. And theory is an essential partner to experimentation and observation, in explaining results, in building the scientific worldview, and in predicting new phenomena. Formal education is essential in today's science, although it is noteworthy how many of the scientists of the past were largely self-educated.

War and peace are certainly important factors. While scientific talent and resources are redirected during wartime, "war is the mother of invention" and has produced technological developments that later had major impacts on science: radar led to radio astronomy, code-breaking led to computers, development of the atomic bomb led to major post-war facilities for the study of the subatomic world, and Sputnik led to a host of satellites and technologies during the Cold War that opened up the entire electromagnetic spectrum for observations of the universe. It became clear that advanced technology would be essential for any wars of the future, and post-war funding for science and technology soared. But peace is obviously essential for the long-term development of science.

In modern science, communication and contact are far more important than they were one or two hundred years ago. The timescales are shorter and communication is faster. Funding has to be secured, so one has to be good at preparing proposals for grants and time using large facilities. Attending conferences and mixing with others is important, both to be up-to-date with the latest developments and to know the other scientists from around the world. It is partly from personal contacts like these, as well as publications, that one's reputation in a field is established.

Some idea of the roles various characteristics and factors have played over the years can be gleaned from reviewing the careers and achievements of a hundred of the most important scientists whose names appeared in the last chapter. This is by no means comprehensive or even a well-defined statistical sample (the very fact that these are some of the most successful scientists in history indicates a strong selection effect), but with these caveats in mind it can perhaps give a rough idea of how science has progressed.

These scientists were all intelligent, strongly self-motivated and determined. They were all driven by pure curiosity (money was not the objective, they were not following orders, and they had no applications in mind). They were all fairly well educated by the standards of the time; most had some formal education, and a few were self-educated.

What is really striking is that almost 90% of the scientists born before 1900 worked and published on their own. While most of them had 'connections' of some sort—helpful contacts, and colleagues around the world who exchanged views and information with them—these scientists had their own ideas, drew their own conclusions, and published on their own. And some of these were solitary indeed. The others worked in partnerships or small groups. A tendency towards collaboration grew slowly with time, particularly over the last century, and considerably less than half of the scientists born after 1900 published on their own. Today, of course, there are some very large teams, even in the thousands.

Half of these hundred scientists were involved in experimental work of some sort (although only about a half of those cases would be considered 'standard textbook experiments'). Over a third did observational work. Technology at some level was of course involved in most of these experiments and observations, and new discoveries resulted from about two-thirds of them. Over half of the scientists did theoretical work. (It should be noted that experiment, observation, discovery and theory are not mutually exclusive). Only a small fraction of the scientists were 'stamp collectors'—those doing the exhausting and meticulous but essential field work of cataloguing all the forms of life and fossils from around the world, and all the stars and galaxies in the sky. And luck featured in the work of less than a fifth of the scientists, although many books have been written on serendipity in science because the stories are so remarkable and fascinating. Finally, only two mistakes appeared in this overview—the famous ones of Einstein and Pauling (see below)—but this is undoubtedly a selection effect, as these hundred scientists are among the most successful of all time.

Incidentally, from a quick glance over the last chapter it is obvious that most of the natural philosophers and scientists had relatively long lives. Those in the Greek, Islamic, medieval and 'modern' (from 1600 to the present) periods had median lifespans of 73, 75, 62, and 75 years respectively. By contrast, the average life expectancy all the way from the Palaeolithic era up to one or two centuries ago was just 25–35 years; the world average in 1900 was 31. In classical Rome and in the UK in 1850 the life expectancy from birth was 20–30 years and 40 years respectively, and even for those who survived infancy up to the age of ten the total life expectancy was still just 48 and 58 years respectively. What can explain the long lives of the natural philosophers and scientists? Scientific output typically peaks in the late 30s if not earlier,[1] so that has nothing to do with the long lifespans. Perhaps it was due to their relatively calm, secluded and reflective lives in fairly good socioeconomic conditions, far removed from the hazards and exertions of normal life. Whatever the reason, we have all benefitted greatly from their long lives in science.

Some of the characteristics and factors in science are elaborated in the sections below, with examples and anecdotes to illustrate how they have been involved in the development of one or more areas of science.

3.1 Curiosity

Curiosity was a major driver—perhaps *the* major driver—of almost all the scientists covered in the history in Chap. 2. From dictionary definitions, it is "an eagerness to know or learn something". So it often leads to an experiment, observation or concept whose outcome or implications are not yet known: "I wonder what would happen if. . ."

Copernicus wondered what the solar system would be like if the Sun was considered to be at the centre rather than the Earth, and worked out the implications. Galileo was curious to see how fast balls of different masses rolled down inclined planes, and to see the heavens through his new telescope. Newton wanted to understand the properties of light. Halley was intrigued as to whether the stars move in the sky. Young was curious to see what would happen if light came through two close parallel slits. Faraday wondered

[1] Jones, Reedy and Weinberg (2014): Age of Scientific Genius.

whether a magnet could induce an electric current. Darwin wanted to know how species evolved. Mendel was curious about heredity in pea plants.

Michelson and Morley were interested to know whether they could detect the aether using experiments on light. Einstein wondered what it would be like to ride on a beam of light, and came up with his famous special theory of relativity. Rutherford was curious to know what would happen if alpha particles were fired into gold foil. Millikan was anxious to know whether he could disprove Einstein's photoelectric effect. Gamow wanted to know what the implications would be if our universe started in a hot and dense phase. Hamilton wondered how an organism could benefit from helping its relatives. Perlmutter, Schmidt, Riess and their colleagues were curious to know how the expansion of the universe evolved. Pure curiosity has always been a major driver in science.

An impressive recent example is that of Nicky Clayton, who started watching the behaviour of western scrub-jays while she was on lunch breaks on the lawns of the University of California at Davis. Unlike most of us who would just casually note the jays and move on, she became very curious indeed about their behavior—so curious that this became a major research programme when she went to Cambridge. If a jay has more food than it needs at the moment, it will cache (store and hide) the excess, sometimes again and again. This requires an excellent memory. Re-caching only happens when other jays are watching or listening, and it is only done by experienced thieves ("it takes a thief to know one"). Re-caching is more likely if the other jay is the dominant member of the group, and unlikely if the other jay is its own partner, with which it would normally share the food. These studies are of considerable importance in the field of animal cognition. They add to the evidence that some animals use their knowledge of the past to plan for the future, and they suggest that these jays may, like humans, have a 'theory of mind'—the ability to ascribe consciousness or 'mind' to others. Casual curiosity on the lawn at Davis led to a major breakthrough in animal behaviour.

Ryszard Maleszka was curious to know how honeybee larvae can become either worker bees (tens of thousands in a hive, which die within several weeks) or queens (one per hive, which lives for several years)—exactly the same genome producing very different outcomes. The only known difference was that worker bees feed some larvae copious amounts of a substance called 'royal jelly', and it is those larvae that can become egg-laying queens. When the honeybee genome was sequenced in 2006 Maleszka saw evidence of methylation (an epigenetic control mechanism), and he wondered whether this is what makes the difference between larvae becoming workers or queens. He and his colleagues found that they could mimic the effects of royal jelly by silencing the methylation of the genome. They could control which larvae develop into queens and which into workers essentially 'at the flick of a switch'; in their experiment they were able to make 72% of the larvae turn into queens, while normally there is only one queen in a hive of tens of thousands of bees. This proved that an epigenetic process is responsible for the difference between workers and queens, and that all the detailed instructions to behave as either a worker or queen are encoded in the *same* genome. It also demonstrated remarkable control over an entire species.

3.2 Imagination

Imagination is almost as prominent as curiosity in the history of science, and both of these characteristics usually appear together in the lives of individual scientists. Again from the dictionary, imagination is "the faculty for creating new ideas".

Newton imagined the forces that govern the world around us, and had the idea that the same law of gravity may govern events both in space and on the Earth. Linnaeus made an important conceptual step in producing the classification system for plants and animals which is still used today. Dalton, studying gases in the early 1800s, came up with an atomic theory that is very similar to the modern view. Darwin assessed vast amounts of information on the world of life, and had the insight that species evolve by natural selection. It is still amazing that this one theory, so simple in concept, explains the huge diversity of life on the planet.

Maxwell had the imagination and mathematics to unify electricity and magnetism in his theory of electromagnetism. Mendeleev and others had the imagination to see how the properties of the elements were related to their atomic weights, and Mendeleev had the insight to leave gaps in this table that would be filled by yet-to-be discovered elements. Planck pondered over blackbody radiation, and had the flash of inspiration that the oscillators producing it may be quantized. Einstein had the 'happy thought' that acceleration and gravity are equivalent, and produced his general theory of relativity.

Bohr imagined how problems with the model of the atom might be solved by assuming quantum-jumping between energy levels. In his PhD thesis Louis de Broglie made the bold suggestion that material particles can be described in terms of waves, introducing the concept of wave-particle duality. After endless discussions with Bohr, Heisenberg came up with the startling 'uncertainty principle', showing that probability reigns in the subatomic world. From the observed motions and distances of galaxies, Lemaître and Hubble both realized that the universe is expanding, and Lemaître drew the conclusion that there must have been a 'beginning'. Waddington wondered how genes could produce differential development in the various types of cells of the body, and introduced the concept of the 'epigenetic landscape'.

Hoyle and colleagues conceived the imaginative 'steady state theory' of the universe as an alternative to the Big Bang theory. Gell-Mann invented the 'quarks' to simplify particle physics at the time when vast numbers of particles were being discovered. Guth thought of 'inflation' as a way of solving the major problems of cosmology at that time. And Mayor and Queloz conceived an ingenious method to discover the first extrasolar planet.

One outstanding example of inspired imagination deserves special mention. Karl von Frisch, an Austrian working in Munich in 1917, discovered a peculiar behavioural pattern in honeybees. When a honeybee returns to the hive, she sometimes does what he called the 'wanzltanz' (waggle dance). She walks in a straight line while waggling her abdomen back and forth, then circles back and repeats the performance again and again. He suspected that this may be some form of symbolic communication, and he was right. The direction of the straight waggling walk is correlated with the direction to a new source of food relative to the direction of the Sun. The length of the waggling walk is related to the distance to the food. And the vigour of the waggle dance is related to the desirability of the food supply. It was astonishing. Von Frisch continued these studies for decades. His careful research initially met with a storm of disbelief and criticism when it was announced, but the waggle dance of honeybees is now widely considered to be the most sophisticated example of non-primate communication known. Von Frisch shared the 1973 Nobel Prize for Physiology or Medicine for this remarkable work.

3.3 Determination

It is impressive how many of these individuals really struggled to get into a position in which they could do science. They were strongly self-motivated, persistent and determined. Galileo himself first joined a monastery and then became a medical student before he finally fell in love with mathematics and dropped out of university to become a tutor of mathematics and natural philosophy. Newton was taken out of school at the age of 16 to manage the family farm. He escaped that fate and managed to obtain a position at Cambridge. He largely ignored the formal curriculum in order to devote himself to science and mathematics. Fortunately he did well enough to be able to stay at Cambridge. He worked as a loner, often with great intensity, and his *Principia* was an absolute *tour de force*. Herschel built his own giant telescopes and spent decades of lonely nights studying the heavens.

Humphry Davy, the son of a farmer, had no formal education beyond a provincial grammar school. He taught himself French, and at the age of 18 read Lavoisier's masterpiece in the original language. He succeeded in becoming an assistant at a research institute in Bristol, where he became well known for his experiments, and at the age of 23 he was made professor of chemistry at the Royal Institution in London. John Dalton also started from humble beginnings, as the son of a weaver. He attended a local school, and when he was just 15 he joined his brother in running a Quaker school. He also gave public lectures, and eventually he was able to support himself as a private tutor in Manchester, giving him time to do his famous science.

Charles Darwin's mother died when he was eight, and he received his basic education at a boarding school. At the age of 16 he was sent to medical school in Edinburgh, but changed to courses in natural history. He was then sent to Cambridge for an education suitable for the clergy, but, being obstinate, he again switched to courses in natural history. Shortly after he finally graduated he received a letter that changed his life—the invitation to join the *Beagle* expedition. Dmitri Mendeleev was the youngest in a family of 14 children in Siberia. His father became blind, and died when Mendeleev was 13. His mother tried to look after the family, and took Mendeleev to Saint Petersburg to receive an education. At first unable to enter university, he became a student teacher before he was able to study chemistry and begin his career in science. Gregor Mendel was born into a poor farming family. He received his basic education at a local school, and did 2 years of university studies before he ran out of money. He became a friar at a monastery, and this finally led to his completing university and returning to the monastery where he did his famous research.

They say that "a scientist with a problem is like a dog with a bone"—he or she is determined, and won't give it up. There have been some remarkable examples of this tenacity over the years.

Albert Einstein displayed remarkable determination over the course of his life. He went to school in Munich, but hated the school's regimen and left to join his family, which had moved to Italy. During this time he wrote a short paper on the state of the aether in a magnetic field—an indication of things to come. At the age of 16 he tried but failed to enrol in the prestigious ETH in Zurich, and finished his secondary schooling at a local Swiss school. He then managed to enter the ETH and completed their math and physics program. He spent two frustrating years searching for a job, and eventually, through a connection (Marcel Grossmann), became an examiner at the Swiss patent office in Bern. There he wrote his famous 1905 papers, the beginning of his stellar career.

Over the following ten years he agonized and struggled with a new project, which was to culminate in one of the greatest scientific triumphs in history. He knew that his special theory of relativity was incomplete, as it was valid only for uniform motion, and he wanted to generalize it to include gravity and acceleration. In 1907 he had what he called "the happiest thought in my life": the equivalence of gravitational and inertial mass, referred to as the 'equivalence principle'. He realized that gravity would bend light rays and cause clocks to go slower. In 1911 some of the basic features of general relativity were beginning to emerge, but it was only in 1912 that he realized that a proper theory would require a non-Euclidean geometry. Fortunately, that was precisely the field of expertise of his old friend Marcel Grossmann, who introduced Einstein to the works of Bernhard Riemann, the great mathematician who in the 1850s developed the geometry of multi-dimensional curved space.

A physical theory of general relativity would require a four-dimensional spacetime, in which "spacetime tells matter how to move, and matter tells spacetime how to curve" (in the words of physicist John Wheeler). Together, Einstein and Grossmann pursued two opposite strategies, one starting from the known requirements of classical physics and the other starting from a purely mathematical formalism. The hope was that these would converge into one correct theory. Einstein and Grossmann ended up with a theory that satisfied the physical requirements, but was not valid in all coordinate systems—that is, it was not generally *covariant*. This was a disappointment, but they felt it was the best that could be done, and in 1913 they published it in what is called their *Entwurf* (outline) paper. In the same year Einstein and Michele Besso used the *Entwurf* field equations to calculate the well-known advance of the perihelion of Mercury; the value they obtained was significantly smaller than that observed. In the same paper they studied rotating frames of reference, and came up with an erroneous result that they thought proved the validity of Mach's principle—that inertia is due to the combined effect of all the matter in the universe rather than the 'absolute space' of Newtonian physics. These incorrect results, in addition to Einstein's nagging concern that the *Entwurf* field equations were not generally covariant, led him to make a renewed effort in the fall of 1915.

This time he started from the mathematical perspective, and demanded general covariance. In a frenzy of effort he was making the pieces come together. In the midst of all this he was to present four papers to the Royal Prussian Academy of Science in November 2015. In the first of these he candidly summarized all the mistakes he had made over the previous few years, explained his renewed search for the covariant field equations, and triumphantly presented his new result. In the second lecture he introduced a small but significant improvement. In the third he showed that his new theory gave the correct result for the advance of the perihelion of Mercury. And in the fourth and final lecture he made a final crucial step; with that he had achieved his goal of a truly covariant theory of general relativity—a huge intellectual achievement and a monumental step in science. Einstein was himself amazed at the power of mathematical formalism to lead to the correct theory. About his own struggles he commented to one colleague "unfortunately I have immortalized my final errors in this battle in the academy papers", to another, "every year [Einstein] retracts what he wrote the year before", and to a third he signed off with the words "contented but kaput".

Einstein, determined to the end, spent the rest of his life in a quixotic effort to produce a unified theory that would combine gravity and electromagnetism. He never succeeded, and even today a unified theory combining all the forces of nature remains the holy grail of modern physics.

Another very determined scientist was the German meteorologist Alfred Wegener, who was impressed by the fact that the shapes of today's continents on a world map seem to fit together like a gigantic jigsaw puzzle. Could it be that the continents had been together in the past, and drifted apart with time? Others had speculated on this before (going all the way back to Aberham Ortelius in the sixteenth century), but Wegener began to take it more seriously. He compared the rock types, geology and fossils on opposite sides of the Atlantic Ocean, and found that they seemed to match to a significant degree. In 1912 he proposed that the continents had indeed drifted apart, and in 1915 he published the book *The Origin of Continents and Oceans.* But his theory was mostly ridiculed, largely for lack of a convincing mechanism. His opponents included professional geologists who saw him as an outsider, and his rather dogmatic style did not help. Nevertheless, he was determined, stuck with his theory, and continued to present more evidence. Unfortunately he died during an expedition to Greenland in 1930 at the age of 50, a few decades before his theory was finally and dramatically confirmed.

In the early 1960s remarkable new evidence supporting the theory of 'continental drift' was pouring in, and by the end of that decade the theory was overwhelmingly confirmed. Much of the new evidence came from studies of the seafloor using echo sounders and magnetometers, which revealed the vast mid-oceanic ridges oozing magma causing parallel zones of reversing magnetic field directions radiating away symmetrically on both sides like zebra stripes—effectively a giant tape recorder of the periodic flips in the Earth's magnetic field. The rate of separation of the Americas from Europe and Africa is a few cm per year—about as fast as fingernails grow, fast enough to be routinely seen in real time with GPS measurements, and enough for the Atlantic Ocean to have formed in just 200 million years, very short on geological timescales. Recently a piece of the Canadian Shield was found in Australia, 11,000 km away, supporting evidence of a supercontinent which started to drift apart over a billion years ago. Wegener is now honoured as the founding father of this major scientific revolution.

Yet another example of determination is that of Leonard Hayflick, investigating some of the properties of animal and human cells beginning in 1958. It had long been 'known' that all normal cells are immortal. When Hayflick was studying tissue cultures, he found to his frustration that the cells he was working on only reproduced a limited number of times—50 or so, in his studies. He struggled to see where he was going wrong, but eventually it occurred to him that perhaps the cells actually do have finite lifetimes. He and his colleague Paul Moorhead then did meticulous experiments in 1961 showing that this is indeed the case, but their paper, which contradicted half a century of dogma, was initially rejected. It took time for their results to be widely accepted, but the discovery was eventually hailed as a major breakthrough and the phenomenon became known as the 'Hayflick limit'. It was a monumental discovery, showing that the cell is central to death as well as to life. It has since been found that the Hayflick limit is related to lengths of non-coding repetitive sequences located at the ends of the DNA strands called telomeres; these protect the DNA until they become depleted through successive replications, when the cell dies. On the other hand there is an enzyme called telomerase that can replenish the telomeres, and make some cells (such as stem cells and cancer cells) immortal. These and subsequent studies have had major repercussions for cell biology and discussions about ageing in general.

A startling implication of the Hayflick limit is that the atoms and molecules in our bodies come and go with time; they are continually being replaced. The time scale depends on the type of tissue—days in some cases and years in others. Over decades virtually all the atoms in our bodies have been replaced. So if you look at a photo taken of yourself taken 20 or 30 years ago, you are looking at a completely different body—none of the atoms and molecules in that person are present in you now! It is *information* that is preserved, not the atoms. The information is that encoded in your DNA and your brain; it is that *information* which defines who you are and gives you a life-long identity.

A researcher in Perth, Western Australia was so determined that he actually risked his life to prove his theory. It had long been believed that deadly stomach and duodenal ulcers were caused by stress, and pharmaceutical companies were making billions of dollars a year from antacid drugs. But in the late 1970s a pathologist in Perth Australia by the name of Robin Warren noticed that small curved bacteria were often found near areas of inflammation. Barry Marshall was a young intern at the same hospital, and became interested in Warren's findings. In 1982 the two of them initiated a study of biopsies from a hundred patients. They found that all of the patients with duodenal ulcers had an excess of the bacteria, which became known as *Helicobacter pylori*, and they then realized that it was those the bacteria that actually caused the ulcers. When they announced their findings there was a storm of disbelief. Their paper was rejected. They realized that their findings threatened not only a $3 billion industry but also the entire field of gastroendoscopy.

In order to definitively prove their case they had to infect animals with these bacteria. They were unable to do this with rats or mice, because *H. pylori* only affects primates. In desperation, knowing of course that he couldn't do the experiment on another human, Marshall cultured some of the bacteria from a patient, mixed them with a broth, and drank it himself! After 5 days he became ill, and 10 days later, after an endoscopy revealed countless bacteria, inflammation and gastritis, he took antibiotics and was cured. His experiment on himself was world famous, and he and Warren were awarded the 2005 Nobel Prize in Physiology or Medicine "for their discovery of the bacterium *Helicobacter pylori* and its role in gastritis and peptic ulcer disease". A dramatic example of a very determined scientist.

3.4 Solitude and Teamwork

It is really quite remarkable how solitary an activity science has been until the last half century. As mentioned above, almost 90% of a sample of some of the most important scientists in the world born before 1900 worked alone. The modern scientific world we live in, with all its technical wonders, is due to just a few hundred key scientists who worked largely alone. Of course they were in contact with others through connections, publications, correspondence and the odd meeting, and many of the scientists worked in a university or institute where daily interactions were possible. But they largely did their work by themselves.

Think of names like Copernicus, Galileo, Newton, Hooke, Herschel, Einstein, Hubble, Zwicky, Boyle, Black, Priestly, Cavendish, Lavoisier, Dalton, Mendeleev, Boltzmann, Planck, de Broglie, Schrödinger, Huygens, Young, Faraday, Maxwell, Lyell, Darwin and Mendel. While they all lived and mixed in society, their scientific work was theirs alone.

Over the years there were cases in which a couple of scientists teamed up, for example Tycho with Kepler as assistant, Ray with Willughby, and Michelson and Morley. But these cases were rare before the last century.

There have also been scientific ventures involving teams of hundreds if not thousands of scientists—but almost all of these were very recent.

The only large collaborations before the last century were the transit of Venus expeditions of 1761 and 1769. These were truly international collaborations—one or two hundred observers from nine or ten countries, who travelled across the globe to make their observations. As these were considered 'missions for all mankind', the participating countries ordered their military to support the observers traveling through their lands, even if the countries were at war with each other. These were remarkable and exceptional endeavours.

In recent times there have been collaborations on vast scales. The large particle accelerators and their detectors are hugely expensive and massively complex, and thousands of scientists have been involved. The Large Hadron Collider is a collaborative effort of the 21 member states of CERN and several other countries, with computing centres in 35 countries and involving thousands of scientists and technicians.

Astronomy has also become 'big science'. The large optical and radio telescopes and their instrumentation are expensive, and large international teams have often been involved not only in making these facilities but also in analysing and publishing the scientific results. Satellites and space missions are also very expensive, and again hundreds of scientists have been involved. The two teams that discovered the acceleration of the expansion of the universe involved dozens of individual scientists, and the collaboration that discovered gravitational waves in 2015 involved over a thousand scientists working in over 80 scientific institutions.

The Human Genome Project involved thousands of individuals working at more than 20 university and institutes in six countries. It remains the world's largest collaborative project in the life sciences.

In looking over the history of science given in Chap. 2, it is clear that large collaborations are a new phenomenon, having arisen in just the last several decades, and they are now a permanent feature of modern science.

3.5 Connections

Sometimes personal connections have helped scientists in various critical ways, for example to get into strategic positions that would enable them to achieve their scientific goals, or to acquire the means or clues to help them in their work.

One very fortunate connection was made in the early 1600s—one that would pave the way for Newton to produce his theory of universal gravitation. When the astronomer Tycho Brahe moved to Prague in 1600 along with his extensive set of accurate astronomical data, he hired the young mathematician Johannes Kepler as his assistant. Shortly thereafter Tycho died, and Kepler had full access to the precious data. With his mathematical background and the Copernican model Kepler was eventually able to derive his three laws of planetary motion, which were critical for Newton's work decades later.

Another connection helped to trigger Newton's work on universal gravitation. Edmund Halley, an acquaintance of Newton's from the Royal Society, made an important visit to Newton in 1684, mentioning the discussion he had had with Robert Hooke and Christopher Wren about whether an inverse square law of attraction could explain Kepler's laws. Within months Newton proved that it could, and within two years he published the *Principia*.

A personal connection was also critical for the career of Charles Darwin. While at Cambridge Darwin studied under John Henslow, the professor of botany, who had a high regard for Darwin. In 1831, when the Admiralty was preparing a surveying expedition to be carried out by the HMS *Beagle* under the command of Captain Robert FitzRoy, Henslow suggested that Darwin be invited to join the expedition to study the natural history and geology of South America. Darwin eagerly accepted, and the rest, as they say, is history.

Another important connection in the history of science was the schoolboy friendship that Einstein developed with Marcel Grossmann. When Einstein was unemployed for two years after graduating from the ETH, it was Grossmann's father-in-law who helped him get a job at the Swiss Patent Office in Bern. And it was while Einstein worked in Bern that he had the brilliant insights that led to his four famous papers in 1905.

When Francis Crick and James Watson were trying to understand the structure of DNA at Cambridge in the early 1950s, they were well aware of the crucial X-ray images being obtained by Rosalind Franklin in the group headed by Maurice Wilkins at King's College in London. At a critical moment in January 1953 they were shown the best of Franklin's images by Wilkins without the knowledge or permission of Franklin. This unauthorized 'connection' gave Crick and Watson one of the final clues to solve the structure of DNA. (Franklin was at that time also close to the solution; unfortunately she died in 1958, four years before the Nobel Prize was awarded for this work—to Crick, Watson and Wilkins.)

3.6 Cross-Pollination

It can be enormously beneficial if knowledge obtained in one field is transferred to another. It can be revolutionary. This can happen in several ways—scientists from two different fields talking with each other, a scientist reading about or moving into other fields, or adopting new technology from another field.

In 1943 the well-known physicist Erwin Schrödinger gave a series of public lectures in Dublin on a topic outside his own field of expertise—life. He wondered how basic physics and chemistry within a cell could explain the secrets of life. On the basis of his lectures he wrote a very influential book in 1944, entitled *What is Life?* This book attracted the attention of many others, amongst them Francis Crick. Crick was a physicist who had worked on wartime projects, and in 1947 he started studying biology, motivated in part by Schrödinger's book. He went to Cambridge, where he met James Watson, and their collaboration led to their discovery of the structure of DNA. In the 1980s Crick moved on to another of his interests: consciousness. He realized that consciousness had long been regarded as a taboo subject by many neuroscientists, and in 1994 he published the book *The Astonishing Hypothesis: The Scientific Search for the Soul* in which he argued that neuroscience had developed to the point at which it should be capable of studying how the brain can produce conscious experiences. His book was inspirational, and the study of consciousness has become an active field of research.

Robert May is another physicist who moved into the life sciences. He is a theoretical physicist who developed an interest in animal population dynamics. He made major contributions to the field of population biology by using his mathematical skills, he helped to develop theoretical ecology, and he applied these tools to studies of disease and biodiversity.

Some areas of mathematics developed long ago have been adopted and used for purposes that were totally unforeseen. The Fourier transform, developed by Joseph Fourier in the early 1800s for his studies of the flow of heat, is a central tool in X-ray crystallography and radio astronomy; the 'fast Fourier transform' is a central feature in the Wi-Fi that we all use today. And Riemann's multidimensional geometry, developed out of pure interest in the mid nineteenth century, was used by Einstein seventy years later for his general theory of relativity.

3.7 Timing

It often happens that the time is 'right' for a particular scientific discovery or development to be made. The science done over the preceding years or decades has prepared the way, and all that is needed is for the scientist(s) of the time to make the crucial step. Sometimes this results in two or more scientists having the same idea at the same time.

Several of the major discoveries outlined in Chap. 2 were made independently by different individuals or teams. The theory of evolution by natural selection was developed independently by Darwin and Wallace, and Darwin got the credit. The Periodic Table was developed independently by four scientists, and Mendeleev got the credit. Mendel's laws of inheritance were discovered independently by five scientists, but in this case Mendel was 30 years ahead of the other four, so he got the credit. The expansion of the universe was proposed by both Georges Lemaître and Edwin Hubble; Hubble had the convincing data, and he got the credit. The structure of DNA was found by both Franklin and Crick and Watson; Crick and Watson were faster in writing up the result, so they got the credit. The microwave background was serendipitously discovered by Penzias and Wilson, but Dicke and his team were at that moment building a telescope specifically to find it, and they would have done so within a matter of months. The acceleration of the expansion of the universe was discovered by both the Perlmutter and the Schmidt and Reiss teams at essentially the same time.

There are many other examples. In the late 1600s the time was right for the universal law of gravitation. Galileo had provided the crucial evidence concerning the law of gravity on the surface of the Earth, and Kepler had provided his three laws of planetary motion. Edmund Halley, Robert Hooke, and Christopher Wren at the Royal Society were discussing the possibility that Kepler's laws could be explained by an inverse square law of attraction. But one may wonder whether anyone but Newton could have put it all together, and in such an overwhelming masterpiece as the *Principia*.

Similarly, in the early 1900s every physicist knew that the speed of light in Maxwell's equations is a constant, and that the Michelson-Morley experiment had found no evidence for an 'aether'. But again, it required Einstein's genius to work through his 'thought experiments' and come up with the special theory of relativity. His general theory of relativity was also a masterpiece, although in this case no one else was close to doing what he did. When he applied it to the universe, it became (and still is) the theoretical framework for cosmology. But others were quick in working out many of its implications.

In the crowded fields of science today it is not surprising that many scientists are worried about the possibility of being 'scooped' by another scientist who is just a bit quicker in publishing the same result.

3.8 Active and Passive Science

We can learn about the world in a variety of ways, all of them important. In experiments we have the huge advantage of being able to vary the parameters (size, mass, time, temperature, speed, electric current, magnetic field, chemical properties etc.), so that the phenomenon under investigation can be studied under a range of conditions. This is typically the case in physics, chemistry and several other fields. The lab conditions are controlled, and we can repeat the experiment as many times as we like, varying parameters to extract maximum information.

But when one has no control over conditions, the only option is to observe and record. This is obviously the case in astronomy, and in much of biology. However, even in astronomy, one can still make hypotheses and use new observations to test them. The observations possible today are remarkably powerful. Thanks to satellites the whole electromagnetic spectrum can be exploited. Surveys of stars and galaxies over huge areas of sky have been of great importance in the history of astronomy. Catalogues of stars go back to ancient times, and in more recent history Tycho Brahe, William and John Herschel and many other astronomers and major national observatories have all made important contributions. It was found that spectra can reveal the chemistry and physics of distant stars and galaxies. The classification of stars made it possible for their evolution to be understood using the Hertzsprung-Russell diagram. The 'collecting' of variable stars by Henrietta Leavitt led to the discovery of Cepheid variables and the distance scale of the universe. Modern catalogues of galaxies of various types at multiple wavelengths have made it possible to study their evolution and ultimately the large-scale structure and evolution of the universe. Our understanding of the universe and its contents is remarkable, considering that it is based on 'mere observation'.

In studying animal behaviour and cognition it is understandable that scientists initially brought animals into the lab and tried to test them with experiments in the laboratory setting as they would a human infant. But it was eventually realized that this was counterproductive for the study of many species. Of course most animals wouldn't perform well in an artificial setting, and we can only really know what they are capable of by observing them in their natural habitats. Fortunately, we have developed sophisticated cameras, sensors and other devices that enable us to observe their 'private lives' at night, in the trees, in their tunnels, underwater, in flight, and at low and high speeds. Some of the most famous early pioneers who studied animals in their natural habitats were Karl von Frisch, Konrad Lorenz and Nikolaas Tinbergen who were jointly awarded the 1973 Nobel Prize in Physiology or Medicine "for their discoveries concerning organization and elicitation of individual and social behaviour patterns". Tinbergen commented that he and the other two Nobel laureates had initially been regarded as 'mere animal watchers', but he strongly promoted the 'watching and wondering' approach to studying animal behavior. The results of these and many others studies over the years have been astonishing, and we can only be impressed by the incredible talents of other animals.

As in astronomy, collecting and classification have also played crucial roles in the life sciences. They date back to the times of the ancient Greeks: Aristotle was the 'father of zoology' and Theophrastus was the 'father of botany'. John Ray and Francis Willughby were the equivalents during the Scientific Revolution, traveling in England and continental Europe and paving the way for the classification scheme of Carl Linnaeus that is still used today. Darwin himself was a collector, and his wide knowledge was obviously essential in his formulation of the theory of evolution by natural selection. Today we know of millions of species: some 300,000 species of plants, 70,000 species of fungi, a million species of insects, 300,000 species of other animals, and far more in the microscopic world. From this wealth of information it became possible to see the relationships and assemble the one 'tree of life' that is central in modern biology.

The collecting and study of fossils dates back thousands of years. Aristotle noted that some fossils in rocks were similar to seashells seen on beaches, suggesting that fossils were once living creatures. The English canal engineer William Smith noted in the early 1800s that rocks of different ages (based on the law of superposition proposed in the late seventeenth century by the Dane Nicolas Steno) contain different groupings and types of fossils, suggesting that these succeeded each other in a regular way. This principal of 'faunal succession' became one of Darwin's key pieces of evidence for evolution. Since Darwin's time the fossil record has been extended back to microfossils and stromatolites more than 3.4 billion years old. In the twentieth century it became possible to use absolute radiometric dating methods to verify and improve the relative ages of fossils, and the fossil record now includes any preserved remains of a once-living creature, from bones, shells and exoskeletons to petrified wood, hair and DNA fragments. Fossils were crucial in establishing Darwin's Theory of evolution, and they remain important today in adding further important details to the tree of life.

Today's 'collections' have become so huge that one talks of 'Big Data'. Huge homogeneous databases of all kinds can be scrutinized in ways never before possible. The data can be 'mined' for very subtle or very rare phenomena that could never be detected in small samples. It is a new way of doing science—a new tool for discovery—and the applications are endless. One involves using computer learning and artificial intelligence to find promising molecules for new drugs from subsets of hundreds of billions of molecules in the vast chemical universe. Another involves using computers to compare digital images of the same fields of hundreds of thousands of stars night after night looking for rare and slight variations in the brightnesses of individual stars—indications of internal physical processes, extrasolar planets or gravitational lenses. A system called DNAStack has been developed by Google to help the global genetics community with the organization and analysis of huge numbers of DNA samples from around the world in order to identify diseases and defects, providing industrial-scale computing power for science. The Large Hadron Collider produces 600 million proton collisions per second of which only a hundred or so are of interest; this requires real-time filtering and rejection of more than 99.9999% of the events. The ancient arts of collecting and analysing now require the most powerful computers and artificial intelligence.

Discovery is a rich source of new knowledge. It may involve exploration, observation, experiment, and often serendipity. Galileo pointed a telescope to the sky and discovered many wonders. Fraunhofer discovered sharp absorption lines in the spectrum of the Sun. Maxwell discovered that a constant in his equations was equal to the speed of light. Lemaître and Hubble discovered the expansion of the universe. Crick and Watson discovered the double helix structure of DNA. Mayor and Queloz discovered the first extrasolar planet.

In 1981 Martin Harwit published a seminal book entitled *Cosmic Discovery*. In it he discussed the concept of a multi-dimensional observational parameter space in the field of astronomy. The parameters include the type of the incoming carrier (photons, cosmic rays, neutrinos or gravitational waves), the wavelength or energy of the carrier that can be detected, the angular, spectral and temporal resolution of the instrument, the polarization properties, the sensitivity, and the time and direction of the observation. Each parameter is an axis in this multidimensional space. The objective is to make observations that are open to the largest possible volume in this observational parameter space, so that the chance of a new discovery is maximized ("the conscious expectation of the unexpected"). Harwit considered a 'discovery' to be a phenomenon that is separated from others by at least a factor of a thousand in any one of the parameters.

This concept has been very useful in thinking about new observations or experiments. In addition to targeting a measurement or phenomenon for a result that is virtually guaranteed (a 'known unknown', in the jargon of risk management and NASA, which would satisfy a conservative review committee), any observation, experiment, new telescope or new instrument should also be open to unexpected new discoveries (the 'unknown unknowns'). This has also been the philosophy justifying small detectors or experiments going 'piggy-back' on large ground-based telescopes while these are carrying out their main missions—the piggy-backers are inexpensive but might provide huge dividends in the form of totally new and unexpected discoveries.

There are many other possible ways of 'planning' for discoveries. For example, when a large area of the sky has been surveyed using both multi-colour imaging and spectroscopy, instead of immediately looking for the unknown one can start by looking for the known. The database of millions of objects can be first examined automatically by computer, which identifies the known classes of objects (stars, galaxies etc.) and puts them aside in a 'known' file; the remaining objects are the interesting ones, to be examined one by one to discover new phenomena.

Discovery is the only way in which many phenomena can be known. Pulsars, for example, could never have been predicted. They are extremely regular flashes of light produced by the rapidly rotating cores of collapsed stars. They are too complicated and involve too many unlikely superimposed layers of physics. Similarly, many phenomena in the complex world of biology can only be known through discovery. It has been said that, in the universe, "whatever is not strictly prohibited is absolutely mandatory". Nature will devise a way of doing it. And only discovery will find it.

3.9 Falsification and Verification

Scientists sometimes just pay lip service to the rules of logic, but science somehow manages to advance nonetheless.

Falsification and verification are closely linked to deduction and induction. Deduction involves drawing a logical conclusion or inferring particular instances from a general law. Induction involves inferring a general law from particular instances.

In his *Principia*, Rule IV of the 'Rules of Reasoning in Philosophy' Newton stated: "In experimental philosophy we are to look upon propositions collected by general induction from phenomena as accurately or very nearly true.... till such time as other phenomena occur, by which they may either be made more accurate, or liable to exceptions"

Hypotheses and theories can be falsified by individual observations or experiments that conflict with them. This is just a matter of deductive logic. The 1930s philosopher Karl Popper was famous for his insistence that a theory must be *falsifiable* in order to be considered 'scientific'; it must 'take risks' and make predictions that can conceivably be proven wrong. This remains a key principle in science today.

On the other hand, theories can never be proven in principle, no matter how many observations or experiments support them. This is the famous 'problem of induction'. The eighteenth century philosopher David Hume argued from the problem of induction that there can be no theory of reality beyond what we directly experience. And Popper's insistence that a theory that has been supported by a great many experiments is no more reliable than one that has been supported by only a few seems to fly in the face of common sense for most of us. A somewhat more nuanced and intuitive approach to the problem of induction is Bayesian inference, named after the eighteenth century English statistician and minister Thomas Bayes. Using Bayes' theorem and 'prior information' it is possible to estimate the probability of a hypothesis being correct, and to update that probability (one way or the other) as more information becomes available. Bayesian statistics are commonly used in modern science. Nevertheless, scientists would agree that ultimately—*in principle*—no theory can ever be regarded as absolutely proven.

But in the 'real' world of science both the steps leading to a hypothesis or theory and the testing of that hypothesis or theory can be complex and varied. Induction and deduction, and verifiability and falsifiability, do not always follow the standard textbook format. Science is pragmatic, and there are exceptions to the principles outlined above. Long-lasting and well-supported theories may not be immediately rejected by a single negative experiment, there are outstanding cases in which a single discovery or development has been considered 'conclusive', apparently flying in the face of the problem of induction, and there are cases in which a hypothesis was accepted as a 'working model' pending the ultimate 'confirmation' which may appear only decades or even centuries later. Real science is pragmatic, and not too hung up on the finer principles of logic.

If a theory is one that has been supported over the years, encompasses a wide variety of observational or experimental facts and is part of a 'web' tying it to a variety of other successful theories, a negative experimental result may lead to a modification of the theory rather than outright rejection. The negative experiment will be examined and repeated and, if it is found to be correct, modifications to the theory will be considered, but any final verdict will not necessarily be immediate. And although a theory can never be considered completely verified in principle, it may become so established that it is essentially considered to be a 'fact', such as the theory that the Sun always comes up tomorrow.

The almost immediate widespread acceptance of the Cosmic Microwave Background (CMB) as conclusive evidence of the Big Bang cosmology in 1964 (described in the last chapter) flies in the face of the problem of induction. This is an outstanding case in which a single discovery *was* actually widely considered to have effectively 'proven' a theory. It was found to have the properties that had been predicted 16 years previously: it is at the predicted temperature and extremely uniform across the sky. This stunning result was so impressive that it was rapidly accepted as overwhelming evidence for the Big Bang theory, and the competing steady state model was dropped by everyone except a few of its most ardent supporters. Actually this 'confirmation' of the Big Bang model was at the same time a 'falsification' of the steady state model, which would produce a superposition of radiation fields from many stars and galaxies, and so could not possibly match the uniformity of the CMB.

The spectacular instant successes of insulin in 1922 and penicillin in 1939 (described below) also did not require an endless chain of supporting experiments. It was immediately obvious that they worked, and massive efforts followed thereafter to produce as much of these 'miracle drugs' as possible, saving millions of lives.

There are many examples in which the 'verification' came decades or even centuries after the hypothesis, while in the meantime the 'hypothesis' had become (part of) the paradigm. Dalton's atomic theory was proposed in 1808, the hugely successful Periodic Table of the elements was proposed in the 1860s on the assumption that atoms exist, Einstein gave his 'proof' of their existence almost 100 years after Dalton proposed his theory, and only now, another 100 years later, have we finally 'seen' individual atoms and even their orbital structure. Darwin's theory of evolution by natural selection was embraced by scientists ever since it was proposed in 1859, even though the mechanism was initially unknown (it has been said that "biology makes no sense without evolution"). But it was only in the last couple of decades that Darwin's theory has been overwhelmingly confirmed by molecular genetics.

Einstein's general theory of relativity was immediately supported by the known advance of the perihelion of Mercury and soon thereafter by the observed bending of light by the Sun. But the other evidence came slowly. The most stunning confirmation, almost exactly one hundred years after Einstein's formulation of his theory, was the detection of gravitational waves in 2015 from two coalescing massive black holes—precisely the complex and intricate signal that his theory predicted. In particle physics the Higgs boson, predicted in 1964, was the last and vital cornerstone of the hugely successful Standard Model, and it was finally detected in 2012 by the Large Hadron Collider, almost 50 years after its prediction; in the intervening years it was 'known' that it must exist, and the Standard Model continued its successes even in the absence of the detection of the Higgs boson.

Newton's theory is still used today in most everyday applications, in spite of the conceptual change brought about by Einstein's relativity. Today, both string theory and the concept of the multiverse have large numbers of adherents, although neither has any experimental or observational support; both have attractive features, and many theorists have been working on them for decades, but at the moment they are just hypotheses.

Scientists are just human, and they can react differently to scientific events. Suppose that, in some field of science, a magnificent 'grand edifice' of theoretical and experimental knowledge has been constructed. Now suppose that, in a flash, it is utterly destroyed by a single overwhelming negative experiment. Some scientists are devastated, and others rejoice. The devastated ones are shocked that such a beautiful construction, created by decades of meticulous theoretical and experimental work, has been destroyed, apparently setting science back by years. The others rejoice that finally we've learned something new—as opposed to all the results that merely confirmed and reconfirmed the edifice. Of course, the edifice is not actually destroyed; it may take years to modify, but the knowledge contained in it is never lost.

3.10 Bandwagons and Paradigms

The 'bandwagon effect' can be good or bad. When a new discovery, a new instrument or promising new theory appears, it is a healthy thing that it attracts attention—scientists are clamouring to follow the new trail or use the new instrument and theorists write papers on the new science that may result, or a promising new theory draws papers that may be critical or supportive and suggests possible new experiments or observations to test the theory. Indeed, there is often a marked spike in publications following a new experimental or observational result, or a provocative new theory. This fast response is beneficial in that it makes science more efficient, especially at a time like the present when so many new facilities and technologies are appearing. Nevertheless, by drawing scientists in from other areas it may be detrimental to those other areas, and it is a matter of historical judgement in each case as to whether, in retrospect, the overall effect was positive for the development of science. A bandwagon can be seen as 'disruptive' in some cases and a leap forward in others.

There have been several recent examples of bandwagons. For example, a flurry of some 500 theoretical papers was prompted by a weak signal seen in LHC data in December 2015, but just months later the 'signal' had disappeared into the noise.

An extraordinarily frenzied event took place in 1987, and is still referred to as the 'Woodstock of Physics' (in reference to the famous 1969 Woodstock Music Festival). It was a marathon all-night session of the American Physical Society on high-temperature superconductivity. Superconductors have huge potential as they can conduct electricity with no resistance, but their use has been limited to extremely low temperatures (below about −240 °C). In the mid-1980s a flurry of papers was appearing on 'high-temperature' superconductors (well above −240 °C), causing great excitement, so a last-minute session on the subject was scheduled. Physicists lined up for hours before the session; 2000 squeezed into the ballroom, many in the aisles, and others watched outside on TV monitors. The session finally ended at 3 am, but many stayed through the night for discussions. The meeting was a media sensation, and many laboratories around the world raced into the field.

Other examples of bandwagons include the discoveries of the first quasars, pulsars and interstellar molecules in the 1960s (many astronomers rushed to the action, and new radiotelescopes were built), the stunning 1995 Hubble Deep Field revealing countless young galaxies in the very distant universe for the first time (huge surveys of distant galaxies involving hundreds of astronomers are now commonplace), and the discovery of the first extrasolar planets in the mid-1990s (hundreds of astronomers rushed into the new field, with the result that over 3700 such extrasolar planets are now known).

A 'paradigm' is the mainstream worldview of a scientific discipline, and a 'paradigm shift' is a significant change in that worldview. A scientist's world is full of paradigms, large and small. The discoveries of the structure of DNA, of quasars, of the Cosmic Microwave Background, of continental drift, of the most distant galaxies, of extrasolar planets, and of the acceleration of the expansion of the universe were all astonishing 'paradigm shifts'; suddenly we understood the mechanism underpinning all life, we became aware of the vast scale of the universe and had direct evidence of its origin, we understood the major geological features on the surface of the planet, we had an entirely new view of the early universe, we became aware of the possibility of countless other worlds capable of hosting life, and we had evidence of the dominant form of mass-energy in our universe. These were all monumental paradigm shifts indeed.

Over the decades scientists make their theories and experiments in the context of the paradigm prevailing in their field. It is completely natural and understandable that scientists have the worldview that has been provided by the science of the past. It could not be otherwise. But it is important that both experimenters and theorists look 'out of the box' from time to time, to think of possible experiments and theories that may lead to new science. This is not always easy, as peer-review panels for funding are naturally conservative. Sometimes the best proposals are those that will provide both conventional results (in part to satisfy the panel) as well as the outside chance of something completely revolutionary.

Even when scientists do a conventional experiment it is always, as an unconscious by-product, a test of the paradigm. Normally the outcome is consistent with the paradigm, and just becomes part of the established body of knowledge in the field. But the situation becomes interesting when the outcome is inconsistent with the paradigm. If the inconsistency is of marginal statistical significance, it might be ignored as a 'noise bump' or the experiment may be repeated—a matter of fine judgement. If it is of high significance, and the scientists are confident of their experiment, the result will certainly be published (suggestions that scientists do not want to shake up the paradigm are totally wrong—every scientist would love to become famous as the one who causes a major change in the direction of science). There will then be questions by others about every detail of the work, digging for mistakes and subtle effects of the instrumentation: "extraordinary claims require extraordinary evidence". Others will repeat the experiment to see if it can be reproduced. There will be a flurry of papers suggesting how the new result can somehow be reconciled with the paradigm. But in the end, if the new result holds up to all scrutiny, the paradigm itself will have to be at least modified if not replaced.

It sometimes happens that one or more scientists refuse to accept the change. They have become 'ensnared' by an outmoded idea that they refuse to give up, even in the face of overwhelming evidence. Some of them relish the feeling that they are right and everyone else is wrong—that they are the renegades and everyone else is passively going along with the mainstream view. This can go on for decades, and the alternative scenarios die only when these scientists do. In the meantime the vast majority of scientists carry on with the new ideas that lead to progress and verified knowledge. There have been several examples of scientists who were crippled by ensnarement over the years.

A somewhat related phenomenon involves the sociology of science. Known as the 'Gold Effect', introduced by the astrophysicist Thomas Gold, it is the bizarre phenomenon in which a scientific idea can percolate and become widely accepted in a field through social interactions in conferences, committees and informal discussions, without any support from empirical evidence. Personalities can play a role, with the most dominant or prestigious individuals having undue influence. The 'club' has simply come to believe in the idea, and this can have a real and distorting influence on the progress of the field, including decisions on proposals and publications. This has been noted particularly in medicine. Obviously such an idea has only a finite lifetime; either it will ultimately be put to the test, or other (confirmed) knowledge will overwhelm it. Nature always has the final say.

3.11 Mistakes

As mentioned above, not many mistakes appear in a brief history such as that given in Chap. 2, but that's undoubtedly a selection effect: what goes into a brief history is the work that actually made progress. Nevertheless, mistakes are certainly made in science as in any other human endeavour, and a few of them are summarized below.

The most famous by far is what Einstein called 'my biggest blunder'. He was applying his new general theory of relativity to the universe in 1917. He knew that the universe was static according to astronomers, but his equations did not allow this—it had to be either expanding or contracting. To solve the problem he added a constant, called the cosmological constant, to his equations, and this made the universe static. Thirteen years later Hubble discovered the expansion of the universe, and Einstein discarded his constant in disgust. It actually seems likely that he made two mistakes, not just one. Inserting the constant was one mistake. But we now know that removing it was also likely a mistake, as the acceleration of the expansion of the universe discovered in 1998 is probably due to the cosmological constant!

Another well-known mistake was Pauling's triple-helix structure for DNA. Linus Pauling, at Caltech, was one of the most famous biochemists of his time. Gradually, the idea that DNA could be the primary genetic material was taking hold, and if so, its structure would be of critical importance. Pauling started to consider it in 1951, but it wasn't until November 1952 that he began working seriously on it. But then he submitted his paper on the subject at the end of December 1952—he had taken just a month to come to a conclusion about the structure of DNA! When Crick and Watson saw Pauling's preprint they knew immediately that it was wrong, and they raced to publish the correct double-helix structure in April 1953. Where did Pauling go so wrong? The X-ray photographs he had were of poor quality. His estimate of the density of the DNA was wrong. He had neglected the critical Chargaff rules. His model conflicted with the acidity of DNA. He had simply rushed to publish a paper containing several fundamental mistakes and based on poor data. A nasty stumble for someone ranked as one of the 20 greatest scientists of all time. Nevertheless, he was awarded the Nobel Prize for Chemistry in 1954 "for his research into the structure of the chemical bond and its application to the elucidation of the structure of complex substances".

A small mistake in The Netherlands led to a very important result. In the early 1940s the astronomer Jan Oort became aware of the fact that the Americans Karl Jansky and Grote Reber had detected radio waves coming from the Milky Way plane of our galaxy. He was especially interested because radio waves can travel unimpeded through interstellar dust, and so can give an unobscured view of the entire galaxy. And if there are any detectable spectral lines at radio wavelengths, they could reveal the motions of distant parts of the galaxy using the Doppler effect. Oort asked a student, Henk van de Hulst, to find out what spectral lines there might be at radio wavelengths. Van de Hulst first considered hydrogen radio recombination lines—the radio equivalents of the well-known hydrogen lines seen at optical wavelengths. Luckily, he overestimated the 'pressure broadening' these lines would experience, and concluded that they were not promising. As a result of this error he persisted and went on to explore other possibilities, and he finally came up with the 'speculative' 21 cm line of neutral hydrogen, which turned out to be far stronger and has played a huge role in radio astronomy.

Another mistake played a major role in the debate between the steady state cosmology and the Big Bang cosmology in the 1950s. According to the steady state model the universe is essentially the same everywhere, while in the Big Bang model the past was very different from the present. Martin Ryle and his colleagues at Cambridge made a survey of radio sources using their new radiotelescope which seemed to show that the number of distant (weak) sources was far greater than the number of close (strong) sources, which, if true, would be evidence of very strong evolution, apparently supporting the Big Bang model. This caused a huge uproar in the debate. But Bernard Mills and Bruce Slee made observations using a different kind of radio telescope in Australia, and found that the actual radio source counts were far less steep. The Cambridge results had been badly contaminated by instrumental effects ('confusion'), and the Cambridge astronomers were careful to correct their errors in their next telescope. (As it turned out, the Big Bang model was correct, but the Cambridge source counts were not!) In spite of this embarrassing episode, Martin Ryle went on to share the 1974 Nobel Prize for Physics.

3.12 False Positives

For an observed 'event' to be claimed as a discovery in the rigorous physical sciences, it must be something that would only happen by chance less than once in 3.5 million times. That is called a 'five-sigma event'. One-sigma events have a one in three probability of being due to chance, two-sigma events have a one in 22 probability, three-sigma events have a one in 370 probability, and five-sigma events have only a one in 3.5 million probability of being due to chance. So one and two-sigma events are usually ignored, three-sigma events are considered interesting, and a five-sigma event is a 'detection'.

Discoveries (first detections) often over-estimate the true magnitude of a signal, which can most easily stand out when it coincides with a positive noise spike. And in any experiment or observation there can be subtle instrumental effects which can masquerade as a signal, so careful experimental control and calibration are of great importance. For these and other reasons the five-sigma criterion, which may seem like over-kill, is strictly maintained in the rigorous physical sciences.

The most famous recent example of a possible signal that disappeared is the two-sigma event that was seen in two independent detectors of the world's largest particle accelerator, the Large Hadron Collider (LHC, briefly described in Chap. 4), in December 2015. In spite of its low significance level, and because of the huge importance of the results from the LHC, some 500 theoretical papers were published in the following months speculating about what it could mean. But by August 2016, with the addition of months of new data, the 'signal' had disappeared below the noise level.

More false positives include those arising from searches for pulses or transient events in astronomy. In 1989 an optical signal with a periodicity at 0.5 ms was detected in the direction of the supernova 1987A, giving rise to more than 50 papers on its implications for neutron-star models, but it was later found to be due to interference from the autoguider on the telescope. In the past decade over 50 transient radio pulses lasting only a few milliseconds have been found coming from different directions of the sky; these are called 'Fast Radio Bursts' (FRBs). In 2010 16 similar pulses were detected using the Parkes radiotelescope in Australia, but these were clearly of terrestrial origin, and they were given the name perytons. It was later found that they were caused by the premature opening of a microwave oven door in a nearby kitchen. But the FRBs themselves are a genuine astronomical phenomenon; they appear to be extragalactic, but their precise origin and physical nature are unclear at the moment.

The WMAP spacecraft studying the Cosmic Microwave Background appeared to indicate in 2001 that the reionization epoch, a major 'phase transition' of the universe, was considerably earlier (when the universe was 60% smaller) than expected on the basis of observations of quasars and galaxies from ground-based telescopes. Hundreds of theoretical papers were written over the next few years explaining the surprising discrepancy in terms of 'early reionization', 'extended reionization', or 'double reionization'. In fact the discrepancy only amounted to one sigma, but that did not deter the theorists. With further data in 2005 the WMAP estimate was reduced to about the expected range, and the most recent result, from the Planck spacecraft, gives a result in reasonable agreement with the evidence from ground-based telescopes. So the discrepancy has evaporated, and the results from the different sources of information have converged (but uncertainties still remain, and the actual detailed history of the reionization epoch has yet to be determined).

A great surprise happened at a very *high* signal-to-noise level in September, 2015: the first detection of gravitational waves, as mentioned in Chap. 2. One would normally expect that, as instruments are improved, any first detection would be marginal, just above the noise level. But almost as soon as the two upgraded LIGO interferometers were turned on, they both detected an incredibly strong 'chirp' (with a signal-to-noise ratio of 24!) extending over two-tenths of a second, with exactly the intricate signature predicted from Einstein's equations. It was obvious in the data immediately. There was no question about it—LIGO had detected two black holes with masses 29 and 36 times the mass of our Sun coalescing in a final crescendo into a single black hole.

It is astonishing that the very first detection was so incredibly strong. But the story is even more intriguing. Because LIGO is so complex and so extremely sensitive (a precision of one ten thousandth of the diameter of a proton), the collaboration had established a secretive and dedicated group whose sole job is to produce 'blind injections'—fake signals added to the data without telling the analysts, to test the detector and the analysis system. These are the ultimate 'false positives'. When the September 2015 signal was detected, it was rapidly confirmed that it was not a false positive—it was very real indeed—a monumental discovery and the birth of an entirely new window on the universe.

Probably the most famous false positive of all time was that of 'cold fusion'. We normally think of nuclear fusion as a process at temperatures of tens of millions of degrees that powers stars and hydrogen bombs. But in 1989 the renowned electrochemist Martin Fleischmann and Stanley Pons at the University of Utah made the startling claim that they had succeeded in fusing hydrogen into helium in the lab, creating heat and traces of nuclear reaction by-products, using a process based on the electrolysis of heavy water (deuterium oxide) on the surface of a palladium electrode. As this happened at room temperature, it was called 'cold fusion'. The announcement of this result happened to come at a time of heightened awareness of environmental issues such as the oil crisis, global warming and the continuing anti-nuclear movement, and the Exxon Valdez oil spill happened just the day after the announcement. So the possibility of a limitless source of clean energy, using only seawater as fuel, caused a media frenzy.

But the announcement rapidly faced a barrage of hostile comments and papers, and the whole episode turned ugly and eventually rather sad. Fleischmann and Pons had rushed to establish their priority with a hasty submission of their publication and a press release. Very quickly many other scientists in several countries announced that they had tried but could not reproduce the claimed results. Some who thought they may have made positive detections later found errors and had to retract their papers. It didn't help that Fleischmann and Pons had not given adequate details of their experiment, and it later turned out that they had not actually detected nuclear reaction by-products after all. Nevertheless, they refused to acknowledge any mistakes in their work, and never retracted their paper. With claims, counterclaims, various interpretations and even lawsuits swirling around, the news media reported "fusion confusion", a "circus", and even "pathological science". Cold fusion eventually became a pariah field, separated from mainstream science, and the remaining cold fusion enthusiasts found it difficult to publish in the major journals. Nevertheless, in view of the possibility, however slim, of a clean and inexpensive source of renewable energy, tens of millions of dollars have been spent on cold fusion research over the years in various countries, and some of this research continues even to the present day. It has been a remarkably complex story involving many aspects of science, scientists and their personalities and interactions; entire books have been written on this saga.

A very different kind of false positive can come from theory rather than experiment and observation. Experimentalists can have a tough time chasing elusive theories. An amusing but frustrating episode played out in the 1970s and 1980s concerning possible structure in the Cosmic Microwave Background (CMB). It was clear from the very discovery of the CMB that its temperature was remarkably constant over the whole sky. But it was realized that there had to be structure in the CMB (small variations in temperature)—the seeds required for the formation of the first stars and galaxies. At first the theories suggested that it may be at the level of a part in a thousand. When new observations reached this level and found no structure, the theories were revised and predicted a part in 10,000. And when the observations reached this level, again the theories were revised, predicting a part in a hundred thousand. Finally, extremely refined measurements using the COBE spacecraft (and later the WMAP and Planck spacecraft) did find structure at this level. That was a relief, as any smaller fluctuations would have been swamped by variations in the galactic foreground emission, and we would never have known about them.

3.13 Disagreements

One of the long-standing disagreements of geological science concerned uniformitarianism vs. catastrophism in the history of the Earth. Catastrophism is the view that the main geographical features of the Earth resulted from gigantic catastrophic events that occurred on short time scales, such as the biblical flood. The main proponent of catastrophism was the French palaeontologist Georges Cuvier, who was motivated by the fossil record rather than theological considerations. Supporting the uniformitarian view were James Hutton and Charles Lyell, who held that the Earth had been shaped by the long-term actions of forces such as uplift, typical volcanoes and earthquakes, erosion and sedimentation that still continue today. There have been many views expressed over the years, but the current view combines both extremes: the Earth's history is one of slow, gradual processes punctuated from time to time by natural catastrophic events.

Within a year of the publication of Darwin's *On the Origin of Species by Means of Natural Selection*, a famous debate took place on 30 June 1860 at the Oxford University Museum, and was known as the 'Great Debate' or the Huxley-Wilberforce Debate. Thomas Huxley, a close friend of Darwin, was championing Darwin's views, while Samuel Wilberforce, Bishop of Oxford, upheld the idea of biblical creation, and railed against the idea that man could be descended from an ape. Both sides claimed to have won, and the matter has remained an issue up to the present, particularly for those of strong religious convictions.

Another Great Debate, called the Shapley-Curtis Debate in astronomy, concerned the nature of the spiral nebulae and the size of the universe. The issue was whether the nebulae were relatively small objects located within our Milky Way galaxy, or whether they were in fact independent galaxies, and therefore large and distant. The debate took place on 26 April 1920 at the Smithsonian Museum of Natural History, and effectively continued in papers published in 1921. Curtis showed that there were more novae in the Andromeda nebula than in the Milky Way, supporting the idea that Andromeda was a separate galaxy with its own rate of nova occurrences. He also pointed out the dark lanes present in other galaxies similar to the dust clouds found in our Milky Way galaxy, and the large Doppler shifts found in the nebulae. In the end, Hubble and others clearly showed that our Milky Way galaxy is only one amongst a great many galaxies in the universe, and the matter was settled, as outlined in Chap. 2.

One of the most famous arguments of all time was the Big Bang cosmology vs. the steady state cosmology in the 1940s to the 1960s. Did the universe have a beginning, or is it eternal? It all started with Hubble's discovery in 1929 that the universe is expanding. The observed redshift-distance relation for galaxies (Hubble's law) implied that the universe was smaller in the past, and extrapolating all the way back it implied a beginning to the universe. There were already theoretical models describing this entire evolution based on Einstein's general theory of relativity, in particular that of the Catholic priest Georges Lemaître, who called the first moment of the universe the Primeval Atom. But most astronomers continued to think that the universe had no beginning, in part because Hubble's early distance scale incorrectly gave an age for the universe that was far less than the well-known age of the Earth.

The issue really started to boil in the late 1940s when George Gamow, Ralph Alpher and Robert Herman were working out the implications of a very compact, hot and dense early phase of the universe, and Fred Hoyle, Hermann Bondi and Thomas Gold proposed the steady state model as an alternative. In this model the universe is eternal, always expanding, but always looking much the same because new matter is continuously being created between the galaxies which are moving apart from one another. One could argue that the creation of this new matter in the space between galaxies is really no more remarkable than the creation of the entire universe at the beginning (a 'Big Bang', as Hoyle once facetiously called it in a BBC interview).

What could discriminate between these two models? Alpher, Bethe and Gamow calculated that the primordial elements (helium, deuterium, lithium and beryllium) could have been produced in the first minutes of the universe, and Alpher and Herman realized that the fading afterglow of the 'Big Bang' should be observable today as a cool bath of radiation present everywhere in the sky.

Observational evidence regarding the two models began to accumulate. The abundance of helium seen in stars and nebulae agreed well with the value predicted from the Big Bang model. The estimated age of the universe was becoming greater with improved observations (Hubble's original estimate was wrong by a factor of ten), so the discrepancy with the age of the Earth disappeared. The new field of radio astronomy was finding some galaxies at large distances (seen as they were a long time ago), that appear quite different from nearby galaxies, implying that galaxies (and the universe) may have changed with time. One survey found that there were many more faint (distant) radio sources than bright (nearby) ones, again suggesting that the universe has evolved; this result caused an uproar, but it turned out to be wrong (see 'Mistakes'). In 1962 the first 'quasar' (quasi-stellar object) was found: a strong radio-emitting object at a high redshift (implying a very large distance) that is as bright as nearby stars. This also caused a storm—how can such a distant object be so bright? And it led some astronomers to suggest that quasar redshifts may not be related to distance at all. In the midst of all this turmoil a single stunning observation almost immediately settled the debate in favour of the Big Bang model: the serendipitous discovery of the Cosmic Microwave Background by Penzias and Wilson in 1964.

3.14 Serendipity

Serendipity (happy chance, good fortune, or luck) has certainly played a role in science over the years, and there are several books[2] on the subject because it is both so important and so intriguing. But it is one thing to stumble across something, and quite another to recognize its significance. As Louis Pasteur once famously said, "Chance favours the prepared mind".

It is not uncommon that a new phenomenon is seen but ignored. Winston Churchill once quipped that "Men occasionally stumble across the truth, but most of them pick themselves up and hurry off as if nothing happened." It takes something special to turn such a stumble into a discovery: a willingness to consider the unexpected, an open mind, a sense of wonder, curiosity, intuition, insight, a determination to follow it up, and creativity in imagining what it may mean and how it fits in. Many important discoveries in science were serendipitous.

Serendipity has been of particular importance in fields such as medicine, as living systems are generally far too complex to predict or to study with simple experiments; their behaviour has to be discovered, so serendipity is often involved.

[2] E.g., Kellermann and Sheets (1983) Serendipitous Discoveries in Radio Astronomy, Roberts (1989) Accidental Discoveries in Science, Meyers (2011) Happy Accidents, and Winters (2016) Accidental Medical Discoveries.

The 'discovery' of a vaccine for smallpox is a prominent example. Smallpox was a terrible scourge for millennia, killing upwards of half a billion people over the course of history. It is a highly infectious fatal disease carried by an airborne virus. But over the years it was realized that survivors of the disease develop a resistance to reinfection. The Chinese and Arabs were among the first to use this clue to provide immunization to healthy people by exposing them to small samples obtained from the scabs or blisters of the victims; the result was usually a mild, temporary form of the disease—and inoculation against any later major exposures. But there was always a significant danger of full-blown smallpox or transfer of the disease if the doses were excessive. By the late 1700s it was realized that milkmaids were generally immune to smallpox. But they were fully exposed to cowpox, a far milder disease, and it was this exposure that gave them resistance to smallpox. In 1796 the English physician Edward Jenner inoculated an 8-year-old using pus from cowpox blisters on the hands of a milkmaid, and then proved beyond doubt that the boy had become immune to smallpox. His thorough researches were a major breakthrough and became well known across Europe, providing the basis for a new and effective vaccine against smallpox. Even Napoleon had all his French troops vaccinated, praising Jenner as "one of greatest benefactors of mankind". Jenner is widely considered to be 'the father of immunology', and in 2002 the BBC named him one of the 100 greatest Britons ever. In 1979 the World Health Organization declared smallpox an eradicated disease, thanks to a huge global public health effort of which this effective vaccine was an important part. But small vials of the smallpox virus have been preserved in laboratories in the U.S. and Russia, and this remains a potential threat, as discussed in the next chapter.

Anaesthesiology also owes its existence to serendipity. It also goes back a long time, involving potions as diverse as the opium poppy in Mesopotamia and cocaine from coca leaves in the Inca civilization, and the term anaesthesia originated (of course) from the Greek word *anaisthetos* ('without sensation'). But the modern development of anaesthesiology began with a series of chance discoveries. In 1799 Humphry Davy became famous for carrying out a series of experiments on the intoxicating properties of nitrous oxide, which he called 'laughing gas', using himself and society friends as guinea pigs. Laughing gas inhalation soon became a fad, featuring in private parties and traveling shows. But Davy also saw the serious side, and wrote a classic book on the subject in 1800. In it, aware that nitrous oxide seemed capable of relieving pain, he made the prescient remark that "it may probably be used with advantage during surgical operations".

Decades passed until the 1840s, when a few American dentists and surgeons started to use nitrous oxide and ether, by then also known to have similar exhilarating effects, for pain-free operations. The word rapidly spread. Another intoxicating gas discovered in the 1830s was chloric ether, which became known as chloroform. It had several advantages over nitrous oxide and ether, and in 1847 the Scot James Simpson promoted its use for childbirth. It was famously used for Queen Victoria in 1853 in the birth of Prince Leopold, and Simpson was knighted in 1866 for his pioneering work. Chloroform became the preferred anaesthetic in the Crimean War and the American Civil War (until that time battlefield operations such as amputations were made without any pain relief whatsoever). From these early 'rustic' developments, the field of anaesthesiology moved on to other more sophisticated drugs, but exactly how they all work is still somewhat of a mystery.

Penicillin was also discovered by chance. Alexander Fleming, Professor of Bacteriology at St. Mary's Hospital in London, returned from a holiday in September 1928, and found something unusual on one of his petri dishes containing colonies of the bacterium *staphylococcus aureus*. There was an area where some mould was growing, surrounded by a zone that was clear of the colonies, as if the mould had secreted something that killed the bacteria. He found that this 'mould secretion' could kill a wide range of harmful bacteria. He had serendipitously discovered penicillin. It proved very difficult to isolate pure penicillin from the 'mould secretion', but this was finally achieved by Howard Florey, Ernst Chain and their colleagues at Oxford. Fleming, Florey and Chain were jointly awarded the Nobel Prize in Physiology or Medicine in 1945 for this life-saving drug.

A similar story was that of Paul Ehrlich, a German physician and scientist working on various areas of medical science in the late nineteenth century. One of his contributions involved the development of staining techniques to identify organisms that cause diseases. In particular, he was trying to stain the tuberculosis *Bacillus*. He tried various dyes for some months, to no avail. One night he left his stained preparations on an unlit stove in his home laboratory to dry overnight. When he returned the next morning, he was shocked to see a fire in the stove, which his housekeeper had lit without noticing the slides. On inspecting the slides through his microscope, he found that the tuberculosis *bacilli* stood out in sharp relief, thanks to the accidental heating, making for easy identification. He (and his housekeeper?) had unwittingly developed an important staining technique, which is still in use today.

As mentioned in Chap. 2, while giving a lecture in in 1820 Hans Christian Ørsted happened to notice that a compass needle was deflected when an electric current from a battery was switched on and off. This small but astonishing serendipitous discovery opened up the entire field of electromagnetism and electrodynamics.

The discovery of X-rays took the world by storm. In 1895 Wilhelm Röntgen was a professor of physics at Würzburg University. He was studying the behaviour of cathode rays emanating from vacuum tubes, and to block out the glow of the light inside the tube he covered the tube completely in black cardboard. Also in his lab was a paper screen painted with barium platinocyanide which would fluoresce when struck by cathode rays; that paper had nothing to do with his current experiment. When he was conducting his experiment in a darkened room, he was astonished to see this screen fluoresce even though it was off to one side well out of the line of fire of the cathode rays. He studied this extraordinary phenomenon, which he called X-rays. His image of his wife's hand showing all the interior bones and her ring immediately became famous. It was soon realized that X-rays are an energetic form of electromagnetism.

Two remarkably similar serendipitous events at Bell Labs in New Jersey led to two of the most important astronomical discoveries of the twentieth century. In the early 1930s Karl Jansky built a radio antenna to investigate any radio noise that might interfere with transatlantic radio voice transmissions. He detected a persistent radio noise over the sky, its intensity rising and falling every day. Eventually he determined that it came from the Milky Way. He had discovered radio emission from our Galaxy, and in doing so he started the new field of radio astronomy. Thirty years later, as described earlier, Arno Penzias and Robert Wilson used a sensitive antenna (at a much higher frequency than Jansky's work) that had been built for satellite communications, this time to accurately measure the emission from radio sources and the sky background. They too found persistent radio noise, which in this case was evenly distributed over the sky—it was the relic radiation from the Big Bang.

In 1967 Jocelyn Bell was a graduate student at Cambridge using a newly built large radio array and the phenomenon of interplanetary scintillation (analogous to the twinkling of stars) to find quasars. The use of interplanetary scintillation required that the telescope be able to detect rapidly changing signals, as fast as a tenth of a second. In the course of her observations she noticed what she called 'a bit of scruff' on one of the chart records. It seemed different from either scintillation or man-made interference, and it seemed to recur on different passes of a particular area of sky. She pointed this out to her supervisor Antony Hewish, who agreed that it should be followed up. By the time he came to the observatory to see for himself, she had recorded the 'scruff' in a faster mode, and it was clear that it was due to very regularly spaced pulses.

They were then able to confirm that the source of these pulses moves overhead with the sky—it is an extraterrestrial pulsating source. One thought that immediately came to mind was that it could be due to signals or beacons of an extraterrestrial civilization. They called it LGM-1 (for 'Little Green Men') and went so far as to consider who to tell about it; the Prime Minister seemed the obvious choice. But after a few weeks the lack of any evidence for (planetary) orbital motion from the precisely spaced pulses seemed to rule that possibility out. With further observations over the sky Bell found three more such sources. Bell and Hewish had discovered pulsars, which, once announced, were quickly determined to be rapidly rotating neutron stars—a spectacular discovery. Hewish shared the 1974 Nobel Prize in Physics with Martin Ryle for their pioneering work in radio astrophysics, citing Hewish's "decisive role in the discovery of pulsars". The omission of Jocelyn Bell from this prize was highly controversial.

In 1979 three astronomers, Dennis Walsh, Bob Carswell and Ray Weymann, were using a large telescope in Arizona to search for quasi-stellar objects (quasars) based on a catalogue of radio sources. They were looking for blue point-like (star-like) objects near the positions of the radio sources. In one case in which they identified such a blue object with a radio source (which was therefore a quasar), they noticed that there was another blue stellar object nearby, just 6 arcs away from the first object. One of the astronomers suggested that it was another quasar that just happened by chance to be exceptionally close to the line of sight of the first, rather than just a blue star. A bet of one dollar was made that it was another quasar. When spectroscopy revealed that it was indeed a quasar, the dollar was paid. But on closer inspection it was found that the spectra of the two quasars are identical. They are in fact two images of the same quasar, seen twice because of 'gravitational lensing' (a phenomenon predicted by Einstein's theory of general relativity). This famous and important discovery was the beginning of the use of gravitational lensing to study the large-scale distribution of mass in the universe. As the second image is not 'another' quasar but the same one seen twice (a far more important discovery), the dollar was paid back.

3.15 The Essence of It All

The complexity outlined above gives an idea of how science really 'happens'. Personalities, social interactions, strategies, technology, serendipity, disagreements and mistakes can all be involved. There is an intricate multi-dimensional web of interleaving observational and experimental results and interconnected hypotheses and theories, and altering one may affect the others. Of course, many of the examples given above are unusual cases, and much of everyday science is not too far from the standard textbook scientific method. But it is clear that it would be impossible to encompass all this variety in a simple algorithm or model. Science is a human activity, and it is therefore endlessly interesting and varied.

Much of the complexity is involved in the establishment of experimental or observational 'facts', and in proposing new hypotheses and theories. No one would disagree that this stage of the scientific process can be very messy (and therefore very interesting). But also the *testing* of hypotheses and theories is not always as simple as the standard textbook model would suggest; in some cases an immediate overwhelming observation or experiment is taken as conclusive evidence, and in other cases hypotheses may last for decades or even centuries before compelling evidence becomes available.

Aha! moments (otherwise known as eureka! moments or epiphanies) have occurred frequently throughout the history of science. Something suddenly becomes clear and obvious. The discovery of the moons of Jupiter and the phases of Venus immediately convinced Galileo of the validity of the heliocentric view of the cosmos. The discovery of the changes in position of some stars relative to others showed Halley that they are not fixed to a celestial sphere. Maxwell's discovery that electromagnetic waves travel at the speed of light immediately showed that light is an electromagnetic phenomenon. The 'discovery' of the Periodic Table gave clarity to chemistry. The discovery of spectral lines in the Sun and other heavenly bodies suddenly made it clear that they are made of the same stuff as we are. The H-R diagram revealed the evolution of stars. The 'discovery' of the velocity-distance relation convinced Lemaître and Hubble that the universe is expanding. The discovery of the structure of DNA immediately implied its functions—an aha! moment for biologists. The Hubble Deep Field suddenly and stunningly revealed the early universe for the first time ever. And the discovery of dark energy, resulting in a flat universe, was certainly an aha! moment for cosmologists. One can try to squeeze each of these into a simple hypothesis-test model, but that's not always how science 'happens'. There is a sudden realization. There is an immediate implication. A whole scenario becomes clear. And events like these are happening at the same time as many other parts of the picture are evolving.

The activities of science are wonderfully complex and varied. They include collecting and classifying, and measuring and determining numbers such as Avagadro's constant, the Hubble constant, the speed of light and the masses of elementary particles. They include the recognition of fossils and their implications for geological timescales, seeing and interpreting the amazing structures of neurons, and Planck's inspired idea of quanta. They include maps of the Earth and of the human genome. They include descriptions of the movements of chromosomes during cell division, Chargaff's rules for the bases of DNA, and Galileo's simple experiments with balls rolling down an inclined plane. There is no simple algorithm to encode all this variety. That is what history is for.

But in the end, and in spite of all this richness and complexity, the essential basic principles still apply. Observations and experiments lead to hypotheses and predictions, and these in turn have to be tested by other observations and experiments. Theories must be falsifiable, and experimental and observational results must be repeatable by others. Theories and experimental or observational results must be rigorously tested, and those that fail must be modified or cast out. Scientific knowledge ultimately has to be grounded by exposure to the real world. It is what works that counts.

4

Science Today

4.1 Exponential Growth

90% of all the scientists who have ever lived are alive today. By contrast, less than 7% of all the people who have ever lived are alive today. It has been estimated that there were a few hundred scientists in the mid-1700s. If the number of scientists had increased at the same rate as the overall population, the number of scientists today would be a few thousand. Instead, according to UNESCO, there are about eight million researchers in the world today. The increase in the number of scientists over the last couple of hundred years is thousands of times the increase in the overall population. The growth rate in the number of scientists since the mid-1700s has been about 4% p.a., corresponding to a doubling time of about 18 years and far faster than the approximately 0.8% p.a. growth rate for the overall population over that period. Currently in China the number of researchers is increasing at the furious rate of 6.6% p.a., while its overall population is growing at just 0.6 % p.a. There is no question that the number of scientists has increased dramatically over the last few hundred years (see Fig. 4.1).

The number of scientists is only one way of monitoring the growth of science. Their effective output appears in publications, and both the number of journals and the number of scientific papers published can be monitored. Derek de Solla Price pioneered investigations of this type in the early 1960s. Initially he used the number of scientific journals, and found that their growth rate was about 5.6% p.a.—a doubling time of 13 years. He then switched to using the number of abstracts of papers published in various fields, and found

P. Shaver, *The Rise of Science*, https://doi.org/10.1007/978-3-319-91812-9_4

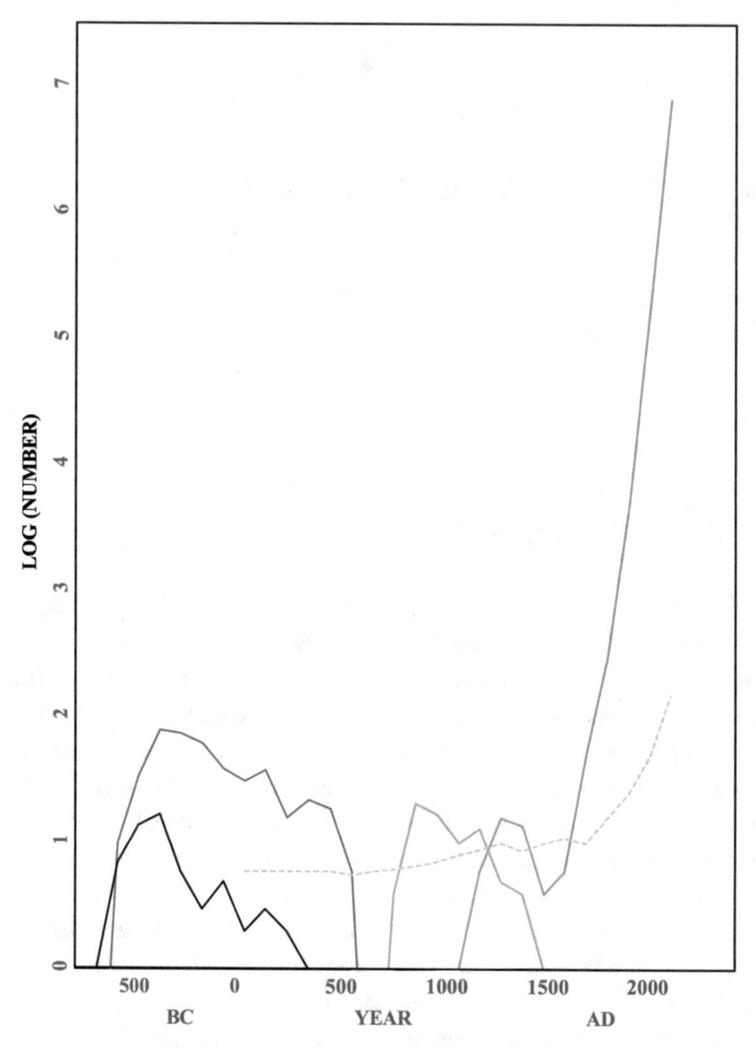

Fig. 4.1 Logarithm of the number of Greek natural philosophers (black line), other Greek philosophers (blue line), Islamic scientists (green line), medieval scientists (brown line), and modern scientists (red line) as a function of time. For comparison, the population of the world (divided by 50 million) is shown as a dashed grey line. Note that the vertical scale covers a factor of ten million. References are as in Figs. 2.1 and 2.2; the growth rates of modern science are from the text in this section

that the overall growth rate was 4.7% p.a., giving a doubling time of 15 years. He reflected on when this rapid growth might reach saturation and level off, and suggested that it may already have started to happen. Now, over 50 years later, we can state with certainty that the rate of growth as measured by Price has not flattened or declined—it has continued unabated up to the present.

Several national and international organizations now regularly monitor the growth of science. A recent paper[1] examined the growth rate for several scientific disciplines from 1907 to 2007, and found that overall the present rate is still about 4.7% p.a., although there are significant differences between fields. Understandably, young disciplines tend to have high growth rates, and more mature disciplines tend to have slower rates. The increasing participation of some developing countries is certainly having an effect. In total, there are now well over 1.5 million refereed papers published each year.

In addition to the numbers of journals and papers, the number of citations of publications can also be used to determine growth rates. A recent study[2] covering the period 1650–2012 estimated that the growth rate was less than 1% p.a. from the mid-1600s to the mid-1700s, then 2–3% p.a. up to the period between the two world wars, and then 8–9% p.a. up to 2012. At 8–9% p.a., global science output doubles in less than a decade!

These estimates are complicated by various factors—very old papers are less likely to be cited, some papers may be counted more than once, and some scientists today publish more than one paper on the same result—but they do at least concur with other studies in showing that science today is growing exponentially. Obviously the growth rate must eventually flatten off well before everyone in the world is a scientist, but at the moment an end is nowhere in sight. As the eight million scientists in the world comprise only 0.1% of the total population, there is room for more growth.

Even more spectacular exponential growth is seen in many of the technologies that power frontline scientific research. In 1965 Gordon Moore, the co-founder of Intel, wrote a paper in which he noted that the number of transistors in an integrated circuit was doubling every 2 years, and he predicted that the trend would continue for the next decade. This became known as 'Moore's law'. In fact it has continued up to the present. The increase since 1965 amounts to a factor of a hundred million, explaining why our modern computers are so powerful and yet so small. The ability to pack ever more transistors into an integrated circuit will soon begin to saturate, however, due to the fact that they are approaching the size of individual atoms. But Moore's law, if it is seen as a measure of the power of computers, may still continue for some time into the future due to better design and more efficient algorithms. And

[1] Larsen and von Ins (2010) The Rate of Growth in Scientific publication and the Decline in Coverage Provided by Science Citation Index.

[2] Bornmann and Mutz (2014) Growth Rates of Modern Science: A Bibliometric Analysis Based on the Number of Publications and Cited references.

then there's the prospect of a totally new technology—quantum computing, which promises vastly greater power for certain applications, as described below.

Something else that has been doubling every 2 years for well over half a century is the beam energy in particle accelerators. M. Stanley Livingston, co-inventor of the cyclotron in the early 1930s, noted this spectacular trend in 1962, and it has since been known as the 'Livingston curve'. The sensitivity of radio telescopes has increased since 1960 with a doubling time of about 3.5 years. Optical telescopes have also increased greatly in sensitivity, due to their increased size and the very powerful digital detectors that became available in the late 1970s, providing almost 100% efficiency compared with the 1% efficiency of photographic plates. The cost of sequencing a human genome has plummeted from $10 million 10 years ago to about $1000 today. And the total information storage capacity available to us, largely in digital form, is doubling every 2 years and measured in zettabytes.

So the number of scientists has been increasingly rapidly, and the efficiency of their tools has also increased very rapidly—together these have provided far more science than ever before.

The dramatic evolution in our scientific horizons is highlighted by the fact that, just over a hundred years ago, we did not know about quantum mechanics, the universe, and the genetic basis of life.

4.2 Curiosity-Driven vs. Goal-Oriented Research

Curiosity-driven sounds frivolous and goal-oriented sounds serious. But actually, it is curiosity-driven research (also known as pure research, basic research or 'blue-skies research') that has given us the most—and the most fundamental—scientific knowledge. This is because it explores the entire natural world, whereas goal-oriented research is focused on specific technical problems. Curiosity-driven research has led to Newton's laws, electromagnetism, Einstein's relativity, quantum mechanics and our knowledge of the evolution and basis of life, all underpinning our modern world as summarized in Chap. 2. In contrast, goal-oriented research is derivative and focused, following predetermined research paths; it applies the fundamentals obtained from pure science to create technical marvels such as the telephone, radio, television, the Internet, the smartphone, the airplane, nuclear power and space travel.

It is therefore understandable that short-term goal-oriented research is mainly done by large corporations seeking financial returns, while long-term curiosity-driven research is left for governments to fund, often through universities. As there is no guaranteed economic return from curiosity-driven research, there is always a struggle to obtain adequate support—even though it is ultimately pure science that underpins our modern world.

Electricity is undoubtedly the most outstanding example of curiosity-driven research leading to world-changing technology. In the early nineteenth century Michael Faraday became a celebrity for his experiments on electricity and magnetism, as summarized in Chap. 2. In retrospect, it is highly amusing to think that anyone could have questioned the value of his research, but they did. When asked about its use he quipped "and what is the use of an infant?" And when the Prime Minister of the time asked the same question Faraday replied "Why sir, there is every probability that you will soon be able to tax it!"

Faraday's work had a monumental impact on world progress and changed the way we live. It transformed cities. The electric clothes washer/dryer helped to emancipate women. Electricity became ubiquitous in homes, offices and industry. It has given us radio, television and the Internet. Faraday would have become very wealthy if he had been able to patent electricity. This story of scientists revolutionizing society with no financial gain has been repeated time and again.

Amazingly, the development of Wi-Fi originated in a curiosity-driven search for exploding black holes in the universe. In 1974 Stephen Hawking showed that hypothetical primordial black holes formed in the early universe may be exploding all around us in the universe today, and Martin Rees then showed that these explosions may best be detected at radio wavelengths. Australian radio astronomer John O'Sullivan and two colleagues working in The Netherlands at the time decided to make a search. One of the problems they faced was the fact that these very short bursts would be blurred in frequency by the interstellar medium. O'Sullivan designed a system to deal with this problem, and observations were made in many directions and at many of the most promising targets. No exploding black holes were found; a paper was duly published in *Nature* in 1978 giving the null result, and the astronomers moved on to other projects.

O'Sullivan, however, remained intrigued by the technical problem he had worked on. Years later, when he returned to Australia, he wondered whether something similar could be used to solve the problem of multipath propagation in wireless computer networking (the many obstacles in a typical office—cables, filing cabinets, walls and furniture—delay and blur the radio signal in much the same way that the interstellar medium does). He developed a computer chip that could perform very fast Fourier transforms to clean up the signal. This was the start of the Wi-Fi revolution. The organization he worked for (Australia's Commonwealth Scientific and Industrial Research

Organization, CSIRO) patented O'Sullivan's invention, and it turned out to be extremely profitable for them, as well as a major revolution for the entire world: there are now many billions of Wi-Fi devices in the world. From exploding black holes to Wi-Fi: a remarkable story of pure science making a major contribution to society.

The laser is another prime example of pure science giving birth to an entire industry. In 1916 Albert Einstein predicted that electromagnetic radiation of an appropriate wavelength can 'stimulate' an excited atom or molecule to fall to a lower energy state and emit more radiation of the same wavelength. The process is called 'stimulated emission'. If several atoms or molecules of the same type are involved and there is a 'population inversion' of the energy states, a cascade of emission can result—an amplification of the original signal. In 1947 stimulated emission was achieved by Willis Lamb and Robert Retherford using hydrogen molecules at Columbia University. Charles Townes, also at Columbia, thought that much stronger amplification could result if he used a population of excited molecules in a resonant cavity with the right dimensions, which could produce a feedback loop. He and his colleagues succeeded in 1954, using ammonia molecules—they built the first 'maser' (microwave amplification by stimulated emission of radiation). The enormously amplified energy was concentrated in an extremely sharp spectral line. In 1958 Townes and Arthur Schawlow proposed a system for optical and infrared wavelengths—a light-emitting maser which was quickly named the 'laser'—and Townes shared the 1964 Nobel Prize in Physics for his work on masers and lasers. Lasers are now everywhere, but Townes, a scientist to the end, moved on to new fields, eventually becoming a professor at Berkeley for the rest of his life and doing pioneering work on several frontier areas of astronomy.

Another revolutionary technology that was born out of pure curiosity is CRISPR genome editing, which makes it possible to insert, delete or edit specific genes or gene sequences in any genome accurately, rapidly and easily. It was a monumental step—it enables us to rewrite the very code of life, potentially putting us in control of our own genetic destiny. Less controversial applications abound, ranging from cures for genetic diseases to improved crops. The American Association for the Advancement of Science named it the breakthrough of the year in 2015.

It all started with the fundamental question of how bacteria fight off viral infections. It was found that the mechanism involved regions of bacterial DNA called 'CRISPRs' *(clustered regularly interspaced short palindromic repeats)*. The regions of DNA sandwiched between these repeating sequences ('CRISPR-associated genes', or simply '*cas* genes') were found to perfectly match the

DNA of known bacterial viruses, and the associated CRISPR RNA molecules precisely direct the Cas proteins to identify, cut up and destroy the viral DNA. That basically solved the fundamental problem of how bacterial adaptive immunity works. But it also raised the provocative question as to whether such a system could be made that can target and cut *any* matching DNA sequence (not just viral DNA). With this in mind, the now known system was modified, simplified and rigorously tested. It worked! A new and extremely powerful technology for precision modification of any sequence of DNA had been produced. It was a revolutionary development. The leading scientists in this work were Jennifer Doudna and her team at the University of California, Berkeley. But many of the crucial steps along the way were made by several other scientists working in different labs around the world—it was a remarkable example of international scientific collaboration at its best.[3]

People have long wondered how geckos can so easily climb vertical and overhanging surfaces. They can hold firmly and then release, with no mark left. Aristotle himself recorded observations of this phenomenon. It turns out to be based on a remarkable system. Geckos have millions of tiny dry, adhesive hairs called setae on their toepads that can provide close contact with rough and smooth surfaces alike. They temporarily bond with surfaces at the molecular level. In addition they have stiff tendons attached to their toepads, which provide high, elastic stiffness in directions in which forces are applied. The result is a powerful adhesive that can be easily removed and leaves no residue. The secret of the gecko's feet was an obvious target for curious researchers, and now that it is understood it has been copied by nanotechnology and applied to adhesives and carrying devices on an industrial scale—for example hospital bandages that can be removed without pain or damage, and devices for adhering to and carrying large glass plates. This new technology was named one of the top five scientific breakthroughs of 2012.

The opposite can sometimes happen. Sometimes goal-oriented research inadvertently produces knowledge of great importance to pure science. Two famous examples, described above, came from Bell Labs—the discovery of radio emission from the Milky Way marking the beginning of radio astronomy in 1933, and the discovery of the all-important Cosmic Microwave Background in 1964; more such examples are given below. Certainly many areas of goal-oriented research overlap with pure research, and in the process of working on practical problems engineers and scientists add to the world's scientific knowledge. The technical challenges of quantum computing, for

[3] Doudna and Sternberg (2017) A Crack in Creation: Gene Editing and the Unthinkable Power to Control Evolution.

example, are at the frontier of quantum physics, and in producing solutions to their problems the computer scientists are adding to the world's knowledge on the behaviour of quantum systems.

It is possible to have the best of both worlds, even within a single institute. A broad goal is set, and the individual scientists or groups of scientists are free to pursue their own curiosity-driven research, following their own routes and whatever serendipitous discoveries they may encounter, within the context of the overall goal. You never know where the next big discovery is going to come from. And serendipity can arise from either pure or applied science. Having scientists from different areas of expertise working within the same institute can increase the probability of advantageous cross-pollination. 'Ecosystems' involving research in pure science, applied science and technology can be extremely beneficial to all. There are obviously many different ways of organizing science around the world.

But in general it can be said that the accumulated scientific knowledge resulting from pure science feeds technology through applied science. Even U.S. President Ronald Reagan once said "although basic research does not begin with a particular practical goal," it is "one of the most practical things government does". The very fundamentals underpinning our understanding of the world come from curiosity-driven research.

4.3 Big and Small Science

Science is done on all scales. There is an understandable tension between big and small science. The limited funds available for science can support large numbers of small projects and individuals that can provide the creativity and serendipity that lead to conceptual breakthroughs, but increasingly expensive facilities and international collaborations are required to make advances in some major fields. The largest and most expensive scientific facilities we have are used to study the largest and smallest scales in the natural world: the universe and the fundamental particles.

To study the universe we would ideally like our telescopes to be above the Earth's atmosphere, which distorts and absorbs most of the electromagnetic radiation coming from space. In fact, we can only see through the atmosphere in two windows: the optical and the radio. To observe the rest of the spectrum—the infrared, ultraviolet, X-ray and gamma-ray wavebands—we have to use satellites and spacecraft. This became possible in the early 1960s, and since then we have launched a great many of them. We have sent spacecraft to explore all the planets in our solar system, and some of the asteroids and comets. These interplanetary missions typically cost in the range $300 million to $3 billion. We have had major satellites and spacecraft studying the universe in all possible wavebands. In total about 80 space telescopes of various sizes have been launched over the years. The major space observatories have a typical cost of $1–2 billion and serve a global user community. The most famous is the Hubble Space Telescope (HST), launched in 1990 and working in the optical and near-infrared bands; its cumulative cost, including building and two decades of operation, is about $10 billion. The next major project is the James Webb Space Telescope (JWST), an $8.8 billion international collaboration led by NASA, to be launched in 2020; it will focus on the infrared wavebands, and should be able to detect the first stars and galaxies in the universe.

From the ground we can observe the universe at radio and optical wavelengths, and we can build much larger telescopes. The current state-of-the-art optical/near infrared telescopes include the European Southern Observatory's Very Large Telescope (VLT) in Chile, the twin Keck telescopes on Mauna Kea in Hawaii, and the twin Gemini telescopes on Mauna Kea and in Chile. These have mirrors of 8–10 m diameter and can support very large instruments. The typical capital cost was in the range $300–600 million. With interferometry and laser guide stars these large telescopes can now compete with the HST in several areas of astronomy. The next-generation telescopes will be ESO's 39-m diameter Extremely Large Telescope (ELT), and the American-led Thirty Metre Telescope (TMT) and 25-m Giant Magellan Telescope (GMT). The ELT, the largest optical/infrared telescope in the world, will cost $1.2 billion, and the operating cost is expected to be about $55 million per year. At radio wavelengths the largest telescopes are the Very Large Array (VLA) in New Mexico, which works primarily at centimetre wavelengths, and the Atacama Large Millimetre/submillimetre Array (ALMA), a collaboration of 20 countries led by Europe, the U.S. and Japan, with a capital cost of $1.5 billion. The next major radio telescope planned is the Square Kilometre Array (SKA), which will work at centimetre and metre wavelengths; it will cost about $1–2 billion, and its two components will be located in Australia and Southern Africa.

The Cherenkov Telescope Array (CTA) will be a very different kind of telescope. It will indirectly detect the highest-energy electromagnetic radiation impinging on the Earth: gamma-rays at energies up to 10^{14} electron volts. Energetic gamma rays initiate atmospheric particle cascades ('air showers') which in turn produce 'Cherenkov light' that can be detected at ultraviolet and optical wavelengths. By using a huge array of telescopes it is possible to determine both the direction and the energy of the incoming primary gamma-rays. This opens up a whole new window on the universe, with a wide range of topics including extreme environments near black holes and other exotic phenomena, the origin of cosmic rays and the frontiers of fundamental physics such as searches for dark matter—and there is always the possibility of totally unexpected discoveries. The CTA will be comprised of 99 telescopes in Chile (hosted by ESO) and 19 telescopes at La Palma, Spain. It is an international collaboration of over 1300 scientists and engineers from more than 200 institutes in 32 countries; the estimated cost is $480 million, and 'first light' is scheduled for 2021.

In contrast to the telescopes mentioned above that have to be at high altitude or above the atmosphere, there is another kind of telescope that has to be as low as possible (underground, under ice, or underwater), and looking down rather than up: neutrino telescopes. Neutrinos can pass almost unimpeded through the entire Earth, so they are very difficult to detect. Their signatures can easily be swamped by cosmic rays entering the atmosphere, so it is best to go as deep as possible and look down at the neutrinos coming up through the far side of the Earth. The giant IceCube Neutrino Observatory at the South Pole is a cubic kilometre grid of detectors embedded in ice 1500–2500 m below the surface. It was completed in 2010 at a cost of $280 million. As IceCube, being in Antarctica, observes the northern hemisphere, another neutrino observatory (KM3NeT, occupying 5–6 km^3 of the Mediterranean) is underway to observe the southern hemisphere. The two will give full-sky coverage—a global neutrino observatory.

Gravitational-wave telescopes have recently opened up another new window on the sky, as mentioned in Chap. 2. The Laser Interferometer Gravitational-Wave Observatory (LIGO), comprised of two large interferometers on opposite sides of the United States, is the largest project ever funded by the U.S. National Science Foundation, with a cumulative cost of upwards of $1 billion. It involves well over a thousand scientists worldwide (the initial detection paper itself had over 1500 co-authors). Virgo is a similar gravitational-wave observatory, located in Italy; it involves 19 laboratories in the six member countries, and several hundred scientists. NASA and the European Space Agency (ESA) have been considering such an interferometer in space, by the name of LISA.

Astronomy advances on a wide range of fronts, and many of them involve far smaller facilities than those mentioned above. Several of the older, well-established observatories still operate with telescopes of 1–5 m diameter and even smaller, and they often host new and innovative small projects, especially if they are equipped with advanced technology. Telescopes of modest size, equipped with state-of-the-art instruments, are used for surveys of large areas of the sky. Wide-angle small telescopes can pinpoint the positions of bursts of light associated with exotic events such as gravitational-wave mergers for the major ground-based observatories. Optical and radio telescopes of modest size dedicated to specific projects exist in locations best suited to their purpose, such as the high Atacama plateau where ALMA is located, and the South Pole. Balloon experiments are often used, as they can reach high altitudes and stay there for weeks at modest cost.

Sometimes important discoveries are made with very small facilities, even the naked eye. On the night of February 23, 1987, the Canadian astronomer Ian Shelton was observing the Large Magellanic Cloud (LMC, a nearby galaxy) with a telescope at the Las Campanas Observatory in Chile, and was surprised to see an unexpected bright star superimposed on the image of the LMC. He went outside and saw it with his own eyes. At about the same time another astronomer at Las Campanas saw it, as did an amateur astronomer in New Zealand. Word quickly spread that there was a supernova in the LMC. It was the closest supernova for over 380 years, bright enough for detailed study. Within hours the most powerful telescopes in the southern hemisphere were trained on this object, which provided copious information about supernovae, including the first detection of neutrinos from a supernova.

The first extrasolar planet orbiting a normal star was discovered in 1995 by Michel Mayor and Didier Queloz using a modest 1.9 m telescope at the Observatoire de Haute-Provence in France. They used a novel method, in which the to-and-fro movement of the star itself caused by the gravitational tug of the orbiting planet was detected. This was one of the most important astronomical discoveries of the past century. Since then over 3700 extrasolar planets have been found. An even smaller facility was used to discover the first case of an extrasolar planet transiting in front of its parent star. David Charbonneau and colleagues made this observation in 1999 with a 10-cm telescope on a parking lot in Boulder Colorado.

Nowadays amateur astronomers take part in the extrasolar planet work by observing transits using small telescopes in their back yards. Amateurs have also discovered many supernovae. The most prolific individual for years was the Reverend Robert Evans in Australia who was able to look at 50–100 galaxies per hour, using his amazing memory of the morphologies of some 1500 galaxies to discover any changes. Thanks to the Internet and modern digital technology 'citizen scientists' are playing an ever-larger role in astronomy, for example examining huge data bases and images for phenomena otherwise overlooked and making 100-year-old papers and images digitally accessible to modern researchers in a project called 'Astronomy Rewind'. Small science can pay big dividends!

The frontier of fundamental physics is at the highest energies, and this requires very large particle accelerators. The biggest is a giant: the Large Hadron Collider (LHC) at CERN, the 21-member European Organization for Nuclear Research, with 2300 staff. It can only be described in superlatives: "the largest, most complex experimental facility ever built, and the largest single machine in the world". It accelerates particles to very high energies in a tunnel 27 km in circumference beneath the France-Switzerland border near Geneva. It is a collaboration with over 10,000 scientists and engineers from over 100 countries, and involves hundreds of universities and laboratories. It can produce 600 million proton collisions per second. The enormous torrent of data is distributed via the Worldwide LHC Computing Grid—the world's largest distributed computing grid—to 170 computing centres in 35 countries. While the LHC collisions can produce the highest man-made temperatures (over a trillion degrees Centigrade) its electromagnets are colder than outer space. In 2012 it achieved its first major goal, the detection of the Higgs boson, and as its energy is further ramped up the search is on for new physics beyond the Standard Model. The cost of the LHC was about $5 billion, and the operations budget is about $1 billion per year. It will provide top-level science for years to come. Its successors for the longer-term future are already being considered—possibly an even larger version of the LHC, or a linear electron-positron collider.

There are of course other facilities in the world with their own major research programmes in fundamental physics. Prominent in the United States are the Brookhaven National Laboratory on Long Island, NY with a staff of about 3000 and annual budget of over $700 million, the SLAC National Accelerator Laboratory near Stanford University with a staff of 1700 and annual budget of about $350 million, and the Fermi National Accelerator Laboratory (Fermilab) near Chicago with a staff of 1800 and annual budget of about $350 million. They each have a variety of accelerators and experiments.

Fundamental physics at the university and institute level is typically split between experimental and theoretical. Many of the experimentalists are involved with the LHC teams or with mid-level laboratories such as those mentioned above. A wide variety of studies are carried out, such as searches for dark matter, precision measurements of fundamental parameters, precision tests of the Standard Model, neutrino oscillations, matter-antimatter asymmetry, nucleon decay and other rare processes, precise measurements of beta-decay, searches for infinitesimal anisotropies in the electron's charge distribution, studies of electron-positron annihilation, searches for charge-parity violation in the neutrino sector, studies of rare decays of mesons, and searches for new physics phenomena generally. The fact that (aside from the

Higgs boson) the LHC has so far found no new particles or other unexpected results gives a new incentive for lab experiments. Perhaps the big breakthrough into 'new physics' beyond the Standard Model will come from a small laboratory rather than the giant LHC.

In the life sciences the biggest project so far was the Human Genome Project (HGP). Its objective was to determine the sequence of all the base pairs in human DNA and map all the genes in the genome both physically and functionally. It was begun in 1990 and completed in 2003 at a cost of $3 billion. The sequencing was done at 20 institutes in six countries. The completion of the HGP was a monumental achievement.

But it was only one (giant) step towards understanding cell biology, which is immensely complex. The expression of the genome is strongly regulated by 'epigenetic' factors, and the ultimate goal is an understanding of the entire 'epigenotype'—all the molecules and interactions, from the genome to the phenotype.

Various new studies are underway or planned. The Roadmap Epigenomics Project is a $200 million U.S. National Institutes of Health initiative started in 2008 to study the regulation of gene activity and expression that is not dependent on gene sequence (epigenomics refers to epigenetic changes across the entire genome). An International Human Epigenome Consortium (IHEC), involving seven member countries, was officially launched in 2010, and aims to produce a thousand reference epigenomes for the international scientific community by 2020. An EpiTwin project between the U.K. and China, another large-scale project in human genetics, will study the differences between identical twins due to epigenetic factors. A new protein catalogue indicating which proteins operate in each of the body's different types of tissues will be of enormous value. A vast international project called ENCODE (the Encyclopedia of DNA Elements) was initiated in 2013 with the objective of identifying every functional element in the human genome. The Human Cell Atlas project, launched in 2016, has the ambitious goal of creating a reference catalogue of all human cells, with their properties and interactions. And the $100 million BRAIN Initiative (Brain Research through Advancing Innovative Neurotechnologies) was begun in the U.S., also in 2013, with the goal of a providing a dynamic understanding of brain function. These large projects will all add to our knowledge of genetic and biological systems, supporting a wide variety of individual studies.

Meanwhile, at the small-science end of the scale, a large number of single-lab projects make progress on many fronts in biology, from bacteria to the complexities of humans. It has been said that creative science works 'bottom-up', not 'top-down'. The large number of independent projects in small science makes possible the creativity, serendipity and conceptual breakthroughs that can lead to major advances; many of the Nobel laureates in these fields worked in small laboratories. How many of today's small-scale projects will revolutionize our understanding of biology and ourselves? The development of CRISPR, described above, beautifully shows how many small independent labs around the world working in parallel and communicating with each other can produce revolutionary science and technology. And, as mentioned in the last section, a large institute with broad goals containing an 'ecosystem' of smaller groups and individuals can sometimes combine the advantages of both big and small science.

Theory was not covered in this overview of facilities and experiments. Theories are of great importance in explaining and making sense of our experiments and observations, and in making predictions. They are part of 'small' science, but only because they just need a pencil, brains and (sometimes) a computer. They are a fundamental part of science. Behind every important fact is an important theory.

4.4 The Support of Science

The world currently spends 1.5 trillion dollars per year on research and development. That is 1.7% of its total gross domestic product (GDP) of $87 trillion. These figures come from UNESCO's Science Report: Towards 2030.

The fraction of GDP spent on research and development varies considerably between countries. In the developed countries it is typically in the range 1.5–3%. In 2013 it was 1.6% in the U.K., 2.2% in France, 2.8% in the U.S., 2.9% in Germany, and 2.4% for the OECD generally. In some countries it was significantly higher: 3.5% in Japan and 4.2% in the Republic of Korea. At the other end of the scale are the least developed countries, with an average of 0.2%.

Why just a few percent? Why not five or ten percent? We spend only a few cents out of every dollar for research and development, while we spend twice as much on entertainment. Somehow, most developed countries around the world seem to have just settled for a research and development budget of only 2–3% of their GDP. And yet all would agree that research and development is the path to economic prosperity.

How is this research and development activity funded and used? In the U.K., France, the U.S. and Germany about 30% of R&D is financed by government (about 20% in Japan and the Republic of Korea). The other 70% is accounted for by business enterprises (about 80% in Japan and the Republic of Korea). In the U.S. most of the funding for basic research comes from the government, and most of the basic research is done by universities.

It is of course not quite so simple. In the U.S. the National Science Foundation (NSF) has provided funding for basic research on a peer-reviewed basis for over half a century, and agencies such as the National Aeronautics and Space Administration (NASA), the National Institutes of Health (NIH), the Department of Energy, Brookhaven, Fermilab, SLAC, the National Radio Astronomy Observatory and many others provide both facilities and support for research. In Europe the European Research Council provides funding for scientific and technological research conducted within the European Union, and there are independent funding agencies and organizations within the individual countries. In addition there are intergovernmental organizations that provide facilities and support research in various fields, such as the European Organization for Nuclear Research (CERN), the European Space Agency (ESA), the European Southern Observatory (ESO) and the European Molecular Biology Laboratory (EMBL). Non-governmental research institutes have been established over the years in several countries, such as the Carnegie Institution for Science in the U.S., the Max-Planck Institutes in Germany, the Tata Institute of Fundamental Research in India, and the Perimeter Institute for Theoretical Physics in Canada. So there are many avenues for scientific research around the world.

In a famous 1945 report to the U.S. president entitled *Science, The Endless Frontier*, Vannevar Bush, then director of the U.S. Office of Scientific Research and Development, expressed his view that basic research is "the pacemaker of technological progress". He wrote that "New products and new processes do not appear full-grown; they are founded on new principles and new conceptions, which in turn are painstakingly developed by research in the purest realms of science". Bush was very influential in the establishment of the National Science Foundation.

The balance between pure and applied science is important. The returns from basic research can be huge, but it takes some courage to support curiosity-driven research, which can seem frivolous, in preference to the sure returns of applied science. So there is a tendency in these days of tight budgets for some developed countries to shift their resources towards goal-oriented research at the expense of pure science. They find it tempting to go for what they hope will be the quick returns of technology and innovation, rather than the long-term benefits of pure research.

Imagine the extreme case in which a developed country stops doing pure research entirely. It then relies on other countries to continue doing pure research, which is openly published and accessible to the entire world. The world's scientific knowledge continues to grow, and the opt-out country just taps into that knowledge to feed its technology. But it loses much more than its own pure research. Its entire capability in fundamental research—the very foundation of its technology—is destroyed; the scientists, with all their curiosity, imagination, creativity and connections to the world's scientific community, have left. The country is reduced to a hollowed-out, third-rate status.

Several decades ago a prominent Indian radio astronomer was asked why his country should support such an apparently irrelevant scientific pursuit when it was in such dire straits economically. He answered that India could not go on purchasing imported machinery indefinitely—it had to develop its own technology, and that required an all-encompassing scientific foundation, including a high level of activity in pure science as well as excellence in science education. India is now a major participant in many of the most important fields of pure research.

4.5 International Collaborations

Large international collaborations are becoming more and more commonplace in science today. There are several reasons. Large experimental or observational facilities such as those described above are very expensive. And as they are at the cutting edge of science, they are in great demand. But they can support only a limited number of users at any one time. The obvious solution is for scientists to form large collaborations and to apply for time as a group. In addition, it is often large collaborations, involving many institutions from different countries, that produce the expensive and complex detectors and instruments for these giant facilities, and these collaborations are rewarded by being given substantial blocks of time when a facility first becomes available; in addition to have access to the first science, these collaborations provide the testing and calibration of the facility plus its detectors and instruments, and this is an important service for the general scientific users at large when they have their turn to use the facility.

There are several other reasons for large international collaborations. The science they want to do is often complex, involving different areas of expertise, and a well-run collaboration of experts in various fields can do far more than a few individual scientists working on their own. Often the scientific objectives require large samples and data reduction, and that means many people working together. And a major reason that large collaborations have arisen in modern science is that they have been made possible by the computer and Internet revolutions.

Some other reasons are less obvious. The European Union encourages and facilitates large Europe-wide collaborations as part of its efforts to bring the different countries ever closer together. And governments generally find international collaborations appealing, both for the efficiencies that such collaborations can provide and for the prestige of being part of something that is important for humanity and transcends national boundaries. Deals can be made, trading time on different facilities in different countries. And for individual scientists in a country that is not a member of the consortium that built the facility, access to that facility can still be achieved by being part of a large worldwide collaboration that includes scientists of the member states.

Countries that are the most open to scientists from other countries and international collaborations have the greatest impact on science. Indeed, this openness is more closely linked to scientific impact than is the amount a country spends on R & D, and scientists that are internationally mobile have significantly higher citation rates than scientists that are not. Today, international projects and collaborations account for over 20% of global spending on scientific research, and up to 50% in some countries. In the U.S. over 60% of the postdocs come from overseas, as did over a third of all U.S.-based Nobel science laureates. European countries rank high in both openness and scientific impact, and the European Union has established the European Research Area to further enhance them; the EU as a bloc is now a strong performer. Countries that turn their backs on scientific mobility and international collaboration do so at their peril.

Scientific collaboration can even provide a bridge between countries that are adversaries. An early example, mentioned in Chap. 2, was the pair of Transit of Venus expeditions of 1761 and 1769, when the British and French governments granted safe passage to the nationals of their rival, as they were "on a mission for all mankind". A more recent example took place during the depths of the Cold War in 1969. Very Long Baseline Interferometry (VLBI), using radio antennas on different continents to work together as a single 'telescope' to give extremely sharp images, was in its infancy. Scientists in the U.S. and the Soviet Union decided to collaborate. Ken Kellermann, of the U.S. National Radio Astronomy Organization, joined his Soviet counterparts at a radio observatory on the Crimean Peninsula for observations in conjunction with radio observatories in the U.S. The atomic clock that he had brought from the U.S. had problems, and he had to fly up to Saint Petersburg to synchronize it with others in Sweden. It occurred to him that to the Soviets on the flight it would have looked the same as if a Russian speaking no English was on a domestic flight from Chicago to Dallas, with an 'atomic' something on the seat beside him. Even more recently, an international synchrotron light source called SESAME has been built in Jordan; the collaboration includes Israel, Iran and the Palestinian Authority.

4.6 Science in Everything

Science is so much a part of our modern lives that we are usually not aware of it or we just take it for granted. When we turn on the lights, put on our clothes, adjust the air conditioning, cook a meal, use the refrigerator, take medications, check the time, drive to work, have a coffee, look at emails, fly to a conference, take a photo, phone a friend, copy a document, buy some groceries, watch television, listen to music, use the dishwasher, have eye surgery or an MRI scan, take antibiotics—we are using the scientific knowledge that has been accumulated over the last few hundred years by dedicated scientists who were driven by pure curiosity.

By far the single most enabling scientific contribution to our modern world is electricity. Our modern lifestyle would collapse without it. Electricity powers a large fraction of modern technology in one way or another. It is the electrical generator that produces this electricity. And yet the generator itself was created out of pure curiosity by the scientist Michael Faraday while he was 'playing' with the properties of electricity and magnetism. The electric motor is just the reverse of a generator: it uses electricity to produce mechanical motion. Electric motors are ubiquitous, driving large machines, robots in factories, electric cars, pumps of all kinds, kitchen appliances, clothes washers and dryers, power drills, vacuum cleaners, DVD players and computer hard drives—all of it originating from Faraday and his curiosity.

The sources of energy to provide power to everything via electricity is a huge issue in our power-hungry society today. Aside from direct sunlight our early ancestors had just simple fires for heat. In the early civilizations they used domesticated oxen and horses to help with heavy labour, water and wind to drive simple machines to grind grain and pump water, and wind in their sails to take them to distant places and facilitate trade. The advent of steam power goes back to the ancient Greeks, but it was only in the seventeenth and eighteenth centuries that innovations by Thomas Newcomen and James Watt gave rise to the coal-fired steam engine that powered the Industrial Revolution and opened up vast new possibilities.

In the late nineteenth century electricity became the currency of energy, and suddenly power could be distributed to virtually anywhere. Electrical power 'grids' extended across entire countries and continents. Generators could convert mechanical motion produced by the flow of water in a hydroelectric plant and the pressure of steam in coal-burning power plants into electricity, which could then be transmitted over long distances to factories and cities. Nuclear power plants, using the process of nuclear fission, added to the energy grids in the 1950s.

But today there are major concerns about the climate impact of fossil-fuel burning power plants and the radioactive contamination potential and 'melt-down' dangers of nuclear plants, and huge efforts are going into cleaner solutions. Giant wind turbines convert the energy of moving air directly into electricity using generators. Vast arrays of solar panels make use of Einstein's photoelectric effect to convert the radiant energy from the Sun directly into electricity, and batteries are used to provide 24-h power. Following Alessandro Volta's invention in 1800, electric batteries became the workhorses for many scientific and technical developments throughout the nineteenth and twentieth centuries. Major efforts are now being made to develop the batteries of the future, which can be used not only with solar arrays but also a vast number of applications, including the electric car.

A big hope for the future is sustainable nuclear fusion—the process that powers the Sun. It is inherently safe and clean, and its fuel, the hydrogen isotopes deuterium and tritium, can be obtained in virtually unlimited quantities from seawater. Another innovation being studied is an 'artificial leaf' capable of converting solar energy, water and carbon dioxide into energy-rich liquid fuels; the process may be ten times more efficient than natural photosynthesis.

Electric lighting is of course ubiquitous in our homes, streets and buildings. The most common forms are incandescent lamps (the familiar lightbulbs), fluorescent lamps (such as neon lights), and modern LED lamps (light-emitting diodes). The strange phenomenon of fluorescence by some rocks had long been known, and by the mid eighteenth century radiant glows had been observed from partially evacuated glass vessels containing mercury and charged by static electricity. In 1802 Humphry Davy produced the first incandescent light by passing an electric current through a thin strip of platinum, and in 1806 he made a much more luminous lamp from an electric arc between two charcoal rods. In 1856 the German glassblower Heinrich Geissler made an evacuated glass tube containing metal electrodes at each end, which produced a luminous glow and ultimately led to the development of commercial fluorescent lamps. By the late 1870s incandescent light bulbs had been independently invented by several researchers, and commercialization followed, most notably by Joseph Swan in England and Thomas Edison in the U.S. Various technical hurdles had to be overcome; filaments made of tungsten and the presence of an inert gas in the bulb gave longer lives. But they are notoriously inefficient: most of the energy is radiated as heat, and less than 5% provides visible light.

The LED solves that problem, using only 10% of the energy required for a comparable incandescent lightbulb, with a far longer lifetime. It is a semiconductor device that emits light when activated by an electric field, in a process called electroluminescence. LED research first took off in the 1950s. The early LEDs emitted in the infrared, but in a seminal 1962 paper Nick Holonyak and S. Bevacqua, working at General Electric's research lab, announced the creation of the first visible-light (red) LED; Holonyak predicted that LEDs would eventually replace the incandescent light bulb, and he was right. In the 1990s blue LEDs were finally developed, which, together with the earlier red and green LEDs could finally produce white light; Isamu Akasaki, Hiroshi Amano and Shuji Nakamura were jointly awarded the 2014 Nobel Prize in Physics for this revolutionary development. The light output of LEDs has increased exponentially (similar to Moore's law), and they are now commercially viable. We are in the midst of another major technological revolution. Mass installations of LED lighting for commercial, public and private use are now common, and the LED lamp market is expected to grow rapidly, from $2 billion in 2014 to $25 billion in 2023—a compound growth rate of 25% p.a.

A microwave oven produces electromagnetic waves in the right frequency range to heat food. Electromagnetic radiation over a huge range of frequencies was predicted by Maxwell's theory of electromagnetism published in 1864, and in 1886 Heinrich Hertz proved that prediction by discovering radio waves. In 1945 the American engineer Percy Spencer accidentally discovered the heating effect of a strong radio (microwave) beam, and the first microwave oven was produced in the same year. Microwaves in the right frequency range (between conventional radio and infrared frequencies) interact with molecules such as those of water, causing them to rotate and interact with other molecules, producing heat.

Our everyday refrigerators and air conditioners produce cooling by making a gas turn into a liquid and back into a gas again in an endless cycle. Three French scientists played key roles in establishing the relevant knowledge. One of them was the seventeenth century French scientist Blaise Pascal, who showed that water pressure increases with depth, and that any external pressure applied to a contained incompressible fluid (such as water) is transmitted equally throughout the fluid (a basic law of hydraulics, known as Pascal's law). In the same century, Guillaume Amontons discovered that the temperature of a fixed volume of gas is proportional to its pressure. In the early nineteenth century Sadi Carnot, the 'father of thermodynamics', studied the most efficient thermodynamic cycles and gave the first formulation of the famous second law of thermodynamics; one of its implications is that heat always flows from hotter to colder bodies, and the temperatures converge to 'thermodynamic equilibrium'. These may seem like esoteric pieces of knowledge, but they are fundamental for many practical applications in our modern world.

In refrigerators electricity drives a compressor, which increases the pressure (and temperature) of the refrigerant in its gaseous state, forcing it out into the condenser coils located at the back of the refrigerator. As these coils are fully exposed to the cooler air of the kitchen, the superheated gas in the coils condenses into its liquid form at high pressure. This saturated liquid then passes through an expansion device and flows into the evaporator coils (located within the refrigerator) within which it can expand, lowering the pressure and therefore the temperature. It becomes a cold liquid, which absorbs the heat from inside the fridge as it evaporates into the gaseous phase. The resulting saturated vapour flows back to the compressor, where the cycle starts all over again. The net result is a forced transfer of heat from the fridge to the kitchen outside, making the fridge cold and the freezer even colder. No moving parts, except for the compressor (and refrigerant). Very clever.

The laser is one of the greatest enabling technological innovations ever. It originated with Einstein's 1916 prediction of stimulated emission from atoms, and was realized by Charles Townes and colleagues in the 1950s as described above. Today, of course, lasers have many applications: barcode scanners, fiber-optic cables, laser surgery, optical disk drives, CD players, laser printers, range and speed measurement, welding, and many others. They are ubiquitous in our modern world.

The Internet and telecommunications revolution have dramatically changed our world. Early forerunners were local-area networks in the 1950s, connecting different users through one computer. The American computer scientist Joseph Licklider proposed a global network in 1960, and led a research group in the U.S. Department of Defence. The first link was established between UCLA and Stanford in 1969—the birth of the Internet. More nodes were rapidly connected, and the Internet grew. The other major development was the invention of the World Wide Web (WWW) by Tim Berners-Lee at CERN in 1989; this made it possible for documents and other resources to be interlinked and accessed by a web browser. The WWW is the main tool used to interact on the Internet. The fact that these developments happened at the same time as the electronics and telecommunications revolutions is no coincidence, and the overall development has been spectacular. The Internet is probably the most important technological development of the past half century. It is estimated that there will be 20 billion devices connected to the Internet in a few years.

Television screens evolved from the cathode-ray tubes being used for fundamental physics research around 1900. For television, beams of electrons were emitted from the cathode in a vacuum tube to light up the phosphorus material on the screen at the other end; the beams were directed by magnetic fields to scan rapidly across the screen to make an image. Modern television screens are quite different. They are thin, flat-screened and use less energy. The most common screens are LCD, plasma, and LED. LCD stands for Liquid Crystal Display. Liquid crystals have remarkable optical properties: an electric field can make a liquid crystal pixel switch between transparent or opaque on command, using its polarization properties. The potential for optical displays is obvious. A plasma screen makes use of small cells of electrically charged ionized gases (called plasmas), and LED screens make use of light-emitting diodes, mentioned above. The commercial market is now dominated by LCD and LED technologies.

Computing devices date back to antiquity, but their modern equivalents are at the frontiers of advanced science. The earliest mathematical aids were simple counting devices. The abacus, which first appeared in Mesopotamia in the third millennium BC, could be used for simple arithmetic tasks. Various other calculating aids, devices and tools for specific purposes were developed over the ensuing millennia (one of the most famous was the Antikythera mechanism, made by the ancient Greeks), but the big step leading to the modern computer was taken in 1833 by Charles Babbage, the "father of the computer". Following his earlier work on mechanical computers, he developed a far more general concept in which punched cards would provide input to a machine containing a logic unit, memory, and control via branching and loops, with output in the form of a printer, plotter or punched cards. It would be a truly general-purpose computer—but his ideas were a century ahead of their time.

The next major step was taken in 1936 by Alan Turing, who proposed a "universal computing machine" that can compute anything by executing instructions and be programmable; this is the basic principle used in modern computers. At that time a fundamental shift was taking place as the old mechanical analogue machines were being replaced by electronic digital computers in the 1930s and 1940s, combining the advantages of the high speed of electronics with the accuracy and reliability of digital (on/off, or binary) signals. In the 1950s the electronic vacuum tubes (valves) were replaced by the far smaller, more efficient, more reliable and longer-lived transistors. A monumental step was taken in 1958 with the introduction of the 'integrated circuit', in which all the components of an electronic circuit are completely integrated on a 'chip' of semiconductor material. Jack Kilby was awarded the 2000 Nobel Prize in Physics for this momentous development, which led directly to the phenomenal electronics and computer revolution of the past half-century.

Now, as mentioned above, the chip components are becoming so small that they are approaching atomic scales (millions can fit on the dot of an 'i'), and new technologies will be required if computing power is to continue its exponential growth. Ideas include new chip designs, alternatives to silicon, improved computer architecture, more efficient algorithms and programming, novel transistors, the use of quantum tunneling, optical techniques, and even emulation of biological brains. Much innovation and creativity will be involved in making better and better computers.

But the big development on the horizon is quantum computing. In a 1981 talk Richard Feynman was musing about how one could possibly simulate the vast complexities of nature (including quantum physics) with a computer, and he suggested that perhaps an immensely powerful 'quantum computer' could do the job. Would such a computer be feasible? The idea percolated for a couple of decades, and has taken off over recent years. The basic unit of information in a classical computer is the bit, which (regardless of its physical realization) has just two possible states, 1 and 0 ('on' and 'off'). The quantum equivalent is the qubit. Quantum computers have potentially two huge advantages over classical computers—two of the weirdest and most esoteric features of quantum mechanics—superposition and entanglement. In *superposition*, a qubit can exist in an indeterminate mixture of the states 1 and 0 at the same time. And in *entanglement,* the quantum states of distinct particles are fixed relative to each other—they are 'bound together', so that what happens to one immediately affects the other, no matter how far apart. Recently, entire clouds of atoms have been successfully entangled. Entanglement is a necessary ingredient of a quantum computer—it unleashes the power of superposition, allowing vast numbers of multiple states to be acted upon in parallel. A 300-qubit machine would represent 2^{300} different strings of 1s and 0s (comparable to the number of atoms in the visible universe), and because the qubits are entangled, all those numbers can be manipulated simultaneously.

Various possibilities are being explored to serve the role of the qubit, including the spins of electrons, nuclei, atoms or molecules, different modes of light, the internal states of trapped ions, quasi-particles called anyons, and the states of small superconducting circuits and currents. Qubits are extremely fragile, and isolation (and in some cases a cryogenic temperature) is required to minimize quantum decoherence. But, while quantum computing is still in its infancy, many developments have taken place and recent progress has been rapid. Several small-scale experimental quantum computers have already been made, the record for maintaining a quantum superposition rose from 2 s in 2012 to 6 h in 2015, and a 72-qubit processor has recently been announced. What applications are foreseen? Quantum computers won't replace classical computers for many mundane general-purpose computing functions, but they will excel in certain areas. Decoding encrypted messages, searching large databases and simulating complex systems (biological molecules, new drugs and materials, and quantum many-body problems) are at the top of the list, and many others are being explored.

The developments on quantum computers are running in parallel with developments on the use of quantum technology more generally, and there are strong connections and synergies. Quantum entanglement is now used in quantum teleportation, with applications in communications. Secure quantum communications have been tested using satellites, and quantum-communication networks are now being set up between multiple cities in various countries. Ultimately there may be a global quantum Internet with quantum-enhanced security using quantum routers and massive quantum cloud computers. Many new technologies based on quantum mechanics are being contemplated, such as cryptography, artificial intelligence, millimetre-precision GPS, optical VLBI, nanotechnology design, super-precise atomic clocks, virtual technology, precision sensors, autonomous vehicles and the Internet of things.

Imagine being able to hear the world's greatest music for the first time ever. For most people this stunning change was due to the gramophone, introduced by Emile Berliner in 1887. The advent of electricity made it all possible. A variety of recording and playback technologies were explored, and eventually disc records emerged as the standard. A recording stylus was used to make a groove in the surface of the disc, and the recorded sound could then be played back when a stylus detected the irregularities in the groove, amplified it and played it through a speaker. Refinements took place over the following half century; the analogue vinyl record became much-beloved, and still is by audiophiles and collectors. In the 1960s tape recorders became popular, and for a while there were three competing types; eventually the lower-quality but more convenient compact audio cassettes won out in the consumer market. In the 1980s records and tapes were largely displaced by digital compact discs (CDs), which dominated the music market for two decades. The disc is etched by a laser beam for recording, and read by another laser beam for playback. A monumental step was the development of all-solid-state digital devices. Having no moving parts they require less power and are lighter, more portable and more robust than disc players. The Apple iPod became iconic in the early 2000s, and now smartphones can carry huge amounts of music in addition to everything else they can do. Another major innovation, due to the telecommunications revolution, is music streaming from the Internet, making available virtually all of the world's music.

The humble telephone originated in the late 1800s. Several individuals are credited with its invention in various forms, but Alexander Graham Bell is generally acknowledged as the inventor of the first practical telephone, and he was the first to patent it, in 1876. In the early telephones the sound waves from a voice caused a diaphragm to vibrate, varying the compression of carbon granules between two metal plates. Electricity passing through the granules was thereby modulated, and the modulated signal could be transmitted through a wire. At the receiving end the modulated electrical signal varied the strength of an electromagnet, which in turn vibrated a diaphragm, creating sound waves. The basic principles haven't changed much over the years, but the technology certainly has. Modern cell phone systems use wireless radio signals to transmit digital information. A small microphone in the cell phone converts the caller's voice into a stream of digital information which is relayed via radio signals to the receiver's phone. We can use cell phones anywhere within reach of radio communication, and for much more than just phone calls—the modern smartphone has thousands of potential 'apps'.

The way we pay for things has been transformed in ways unimaginable just several decades ago. Trade originated over 10,000 years ago in the form of barter—the direct exchange of goods. Eventually standards and currencies were invented to serve as go-betweens and facilitate trade, and the first coins appeared in the seventh century BC. Over all the years since then things didn't change much. Paper currency was introduced by the Chinese two thousand years ago, and banks and the concept of deposits emerged 500 years ago. But even just 50 years ago, for most people finance still meant deposits, bank accounts, cash and cheques. Then credit cards were gradually introduced—initially a low-tech kind of payment based on signatures and manual processing; the process was finally computerized in the 1970s. Since then there have been major technological changes. The magnetic strip, developed from computer data storage technology, was introduced on credit cards in the 1970s. The electronic integrated circuit chip was added in the 1980s and 1990s. Security holograms were added to further deter counterfeiters. And contactless systems using radio induction technologies that do not require physical contact between a card and reader have been introduced over the last twenty years, enabling instant 'tap' payments up to a certain amount. These and other technologies are starting to reduce the need for cash. In China this change is rapidly taking place today; over just the last 3 years almost everyone in the major Chinese cities has switched from cash to smartphones for making payments. Will cash become obsolete? Meanwhile Internet currencies such as Bitcoin are appearing, and may further change the landscape.

The number of photos taken in just one year is now several trillion—far more than all the photos ever taken on film over the entire history of photography. The road to photography was a long one. The pinhole camera (a light-proof box with a tiny hole at one end that gives an inverted image at the other end) was known to the ancient Chinese and Greeks well over two thousand years ago. By the seventeenth century the pinhole had been replaced by a lens, and the camera was used as a drawing aid. In 1614 Angelo Sala found that powdered silver nitrate is blackened by the Sun. In the early 1800s Thomas Wedgewood, Humphry Davy and Nicéphore Niépce all had some success in capturing crude, impermanent shadow images using silver nitrate. Niépce went on to use bitumen as a coating on pewter, which hardened on exposure to light; the result was the oldest surviving photograph. Following the untimely death of Niépce, his partner Louis Daguerre carried on, using silver-based processes, and in 1839 finally announced the first practical photographic process to the French Academy of Sciences and the world. Many developments rapidly followed, the most important being that of photographic film by the American George Eastman in 1884, and the Eastman Kodak Brownie camera made photography available to the mass-market in 1901.

A century later, photography was totally revolutionized by a remarkable 'disruptive technology'. Electronic arrays can be used instead of film to detect light. One of the many marvels produced by the electronics revolution of the past half century is the charge-coupled device (CCD), invented in 1969 by Willard Boyle and George Smith at Bell Labs. An image projected onto an array of very sensitive capacitors gives each element an electrical charge corresponding to the light it receives. A control circuit shifts the electrical charge from one capacitor to the next, and ultimately to a charge amplifier and digitizer. In this way the image is captured and 'read out' into a digital form, which can then be stored in a memory chip or computer. The images produced by such digital detectors have huge advantages compared to photographic film. They are far more sensitive, they can be viewed instantly (compared to the hours, days or weeks for film), they cost nothing, and they can easily be manipulated, copied, and sent around the world in seconds. In retrospect it is clear that digital would rapidly eclipse film. The ultimate irony is that, while Kodak invented digital photography in 1975 and introduced the first commercial digital cameras in 1995, it stuck with film as its core business and went bankrupt in 2012. In fact, digital photography almost completely replaced film within a decade—probably the fastest 'technological disruption' in history.

We may not always think of it that way, but the modern automobile is an engineering marvel, involving a host of diverse technologies. At its core is the internal combustion engine. It seems hard to imagine that the smooth, silent motion of a modern car is actually the result of thousands of explosions taking place every minute in several cylinders that convert that energy into the rotation that drives the wheels. Just as important as the propulsion mechanism is the ability to stop; this also requires great force, and it is provided by the hydraulic braking system, based on Pascal's law and the conversion of kinetic energy into heat. We have come a long way since the first automobile was developed by Karl Benz in 1885. Modern cars are designed using computer-aided design systems, and can be dynamically tested in sophisticated computer simulations before even a prototype is made. Assembly lines are highly automated. The typical car today has dozens of 'computers' (central processing units) controlling many of its functions, and that number will rise. In fact the car is beginning to resemble the jet airplane in technological sophistication—perhaps not surprising, as cars operate in far more crowded and complex environments than do airplanes. And the requirements will become far greater as the era of self-driving cars approaches.

When you travel somewhere, you can follow your own position in real time in a car or on your cell phone. This is made possible by GPS (the Global Positioning System), completed by the U.S. Department of Defence in 1995. It has been made freely available to civilian and commercial users around the world, and other countries and groups of countries have now also made their own systems. The GPS system involves at least 24 satellites operating at any one time. It is based on a precise knowledge of time and the accurately known positions of the satellites. The satellites carry very stable atomic clocks that are synchronized to each other and to ground clocks. They continuously transmit their current time and position to any receiver on the ground; at least four satellites must be in view of the receiver for it to be able to compute its position (to an accuracy of a few metres). Obviously many areas of scientific knowledge have come together to make this possible, including the physics of atomic clocks, an understanding of radio waves, and the technologies to launch satellites into orbit. The most esoteric physics involves Einstein's theories of special and general relativity: the special theory predicts that the satellite clocks will fall *behind* clocks on the ground due to their speeds, and the general theory predicts that the satellite clocks will get *ahead* of those on the ground due to the smaller gravitational field they experience because of their height above the Earth. Both effects have to be computed, or else there would be errors in the GPS positions accumulating at about 10 km per day! So, while Einstein's theories may normally seem to be far removed from daily life, without them the GPS system would be useless.

Knowing the time used to be based on the movements of the celestial bodies and later on the swing of a pendulum, but today it is based on 400 atomic clocks kept in national laboratories around the world. These clocks are continuously compared with each other using GPS and other satellite communication networks, and 'the' time is a weighted average of all of them. The clocks are based on the precise frequencies that atoms emit when they change energy levels; the atomic physics involved is that developed by the scientists Neils Bohr, Albert Einstein and others a hundred years ago. The international network of clocks is synchronized to an accuracy of a billionth of a second per day. Even more accurate clocks are now being developed that will not gain or lose a second over the entire lifetime of the universe, 13.8 billion years.

How do airplanes fly? Two major requirements are speed and the shape of the wings. The speed is provided by the jet engines, which use Newton's third law. The fuel burning in the engines produces hot gas that is expelled out of the rear of the engines, and the reaction propels the plane forward. The shape of an airplane's wings (called an airfoil) gives it lift. The top is curved, while the bottom is flat. This forces the air passing over the top to move faster than the air flowing below; according to a principle named after the scientist Daniel Bernoulli who first published it in 1738, the faster-moving air above has a lower pressure than the slower-moving air below, resulting in an upward force.

Even greater lift can be provided by the 'angle of attack' of the wings, increased when the pilot pulls up the nose of the aircraft so that more of the underside of the wings encounters the airflow. Planes can even fly upside-down using this principle. Helicopters rely on rotating blades for lift; the angle of the blades can be controlled by the pilot as required. The helicopter was the earliest known 'flying machine'. By 400 BC Chinese children were already playing with bamboo flying toys comprised of blades attached to a stick that could be rolled to spin the device, creating lift. In the fifteenth century Leonardo da Vinci made designs for flying machines; he considered both vertical and horizontal flight, and he realized that manpower alone was not sufficient for sustained horizontal flight. Despite a great many attempts over the centuries, it was only when the internal combustion engine became available that sustained manned heavier-than-air flight was achieved—by Orville and Wilbur Wright in 1903 for horizontal flight, and by Louis and Jacques Breguet in 1907 for vertical flight. Developments since then, of course, have been meteoric.

A relatively unsung hero for many things that move, including ships, cars, planes, satellites and spacecraft, is a device known as a gyroscope. We are all familiar with spinning tops as toys for children; they originated independently in all the early cultures of the world. Spinning tops continue spinning because of inertia: Newton's first law of motion states that the motion of a body will persist unless an external force is applied to it. In 1852 Léon Foucault used a gyroscope (as well as his famous pendulum) to demonstrate the rotation of the

Earth on its axis. A classical precision gyroscope is a rapidly spinning wheel mounted on two or three gimbals. The wheel maintains its orientation regardless of the motion of the outer frame, so sensors can detect any accelerations. Miniaturization of the gyroscope has gone to extremes in recent decades. MEMS (Micro Electromechanical Systems) are micrometre-size mechanical devices built into electronic chips; they use vibrating elements rather than spinning wheels to detect accelerations and orientation, as do some insects such as flies, and are called vibrating structure gyroscopes. So the technology of gyroscopes has gone from spinning tops to gimballed wheels to microelectronics. They have a huge number of applications today, including ship, aircraft and spacecraft stabilization and navigation, guidance systems in missiles, Segways, underground tunnelling, activation of automobile air bags, and image stabilization in cameras and personal smartphones.

Particle accelerators were originally developed many decades ago for research in particle physics, but they are now also used for a variety of other purposes. There are currently over 30,000 accelerators in operation around the world. Cyclotrons are used to produce radioisotopes for medicine. Synchrotrons can produce extremely intense beams of electromagnetic radiation (EMR) which are used for a wide variety of purposes: biology (molecular structure), medical research (microbiology, cancer radiation therapy), chemistry (composition analysis), environmental sciences (toxicology, clean technologies), agriculture (plant genomics, soil studies), minerals exploration (analysis of core samples, studies of ores), materials (nanostructured materials, polymers, electronic and magnetic materials), engineering (imaging of defects in structures, imaging of industrial processes), forensics (identification of extremely small samples), and cultural heritage studies (non-destructive techniques for palaeontology, archaeology and art history). The ultimate EMR source is the new 3.4 km-long European X-ray free-electron laser (XFEL) that produces X-ray pulses a billion times brighter than conventional synchrotrons can achieve. The super-fast time resolution makes it possible to follow high-speed biology at the atomic level.

Some fasteners and adhesives that we commonly use today were inspired by observations of nature. The idea for Velcro occurred to a Swiss engineer who was out walking his dog in the woods in 1941 and was intrigued by the burrs that became attached to his clothes. He thought of a way of reproducing the mechanism using many tiny hooks on a piece of material that could catch onto another, receptive, piece of material, and later be pulled away with ease; he patented Velcro in 1955. Today it is used in a wide variety of applications. And, as mentioned above, the amazing ability of geckos to walk up smooth vertical surfaces captured the attention of many people over the ages, and when the mechanism was understood it was quickly copied and used for medical bandages and devices for adhering to and carrying large panes of glass leaving no residue after release. But there is one substance that even geckos cannot adhere to: Teflon, which was discovered by accident by a DuPont scientist in 1938 while looking for non-toxic refrigerants. It is one of the most slippery substances known and repels other matter, including the feet of geckos. It is best known for its use in non-stick pans, but it is also used in a host of other applications. At the other extreme, one of the most sticky substances known (cyanoacrylate) was first discovered by accident in 1942 by Goodrich scientists who were trying to make clear plastic gun sights for the war effort. It was rediscovered by Eastman Kodak researchers in 1951, and this 'superglue' rapidly became famous and widely used.

Another form of glue was recently inspired by nature: *slug slime* turns out to have remarkable properties that are highly advantageous for medical adhesives. Certain slugs ooze a sticky substance to ward off predators. An international team of scientists has studied the properties of this slime. It is a matrix called a hydrogel, a web of starchy chains of molecules that is mostly water. Its remarkable adhesion properties result from electrostatic interactions, covalent bonds and interpenetration. Because it is stretchy, it can move with body tissues while maintaining its grip, and leaves less of a scar than normal surgical techniques. It can even stick to wet surfaces and is effective in the presence of blood. It can be injected, it can be attached to a beating heart, it will stick strongly to skin, arteries and internal organs, and it may eventually replace stitches in wound repair. A major contribution to medicine from the humble slug.

Science of course plays a huge role in medicine today. But medicine was late in becoming scientific. The ancient Greek humours were very much a part of medical thinking into the early nineteenth century, and the long tradition of bloodletting only died off towards the middle of that century. By that time the public health movement, with its organized and common-sense approaches, was starting to have a significant positive effect on general health and longevity. But it was only in the late 1800s that meticulous scientific research discovered the causes of several diseases. The microscope, long ignored for almost two centuries, finally revealed the bacteria causing diseases such as anthrax, cholera and tuberculosis. Louis Pasteur and Robert Koch were the pioneers in this field, producing the 'golden age of bacteriology'. In the 1880s Pasteur produced vaccines against anthrax and rabies with great success, and the diphtheria antitoxin followed. The cutting edge of medicine had moved into the laboratory.

Two of the outstanding success stories in scientific medicine are those of insulin and penicillin. The road to insulin was a long one. Diabetes had been known as a disease for millennia (there are over 400 million cases today, and diabetes is the seventh leading cause of death in the Western world), but it was only over the decades from the 1860s to the 1920s that the pieces of the puzzle were gradually put in place. In 1869 microscopic studies of the pancreas revealed unusual clumps of matter, later called islets. In 1889 it was found that the removal of the pancreas in dogs caused diabetes, and it was soon suggested that the islets may play a regulatory role in digestion. In 1901 it became clear that diabetes is caused by the destruction of the islets. There followed two decades of attempts to isolate whatever it is the islets secrete. In 1921 Frederick Banting at the University of Toronto thought of a way of obtaining a pure extract from the islets before they could be destroyed by the enzymes of the pancreas, and he and his colleagues succeeded in producing the active hormone insulin (*insula*, Latin for island). The first injection of insulin was given to a 14-year old boy who lay dying in 1922, and he rapidly recovered. Equally dramatic were cases of patients in diabetic comas being revived by the administration of insulin and glucose. These were spectacular achievements, which would save and improve the lives of millions of people.

Penicillin has been called 'the wonder drug of all time', a 'miracle drug', and 'the most important discovery of the past millennium'. It produced a revolution in medicine. Its serendipitous discovery by the Scot Alexander Fleming in 1928 was briefly mentioned in the previous chapter. Fleming had the 'prepared mind' and discipline to investigate and follow it up in detail. But he was a poor communicator, and his momentous discovery attracted little attention until 1939 when an Oxford team led by the Australian Howard Florey purified penicillin and showed that it cured infected mice and some human patients. After this, penicillin production was ramped up for the war effort, especially in the U.S., and by the end of the war over 600 billion units were being produced annually. It cured infections, gangrene and pneumonia in days, and saved the lives of hundreds of thousands of wounded soldiers. But this was only the beginning of the scientific story. Penicillin was in wide use following the Second World War, and in those heady days it was predicted that medical science would completely overcome all of man's microscopic enemies within a generation. But the ability of microorganisms to evolve and become resistant to penicillin came as a surprise. To cope with the new bacterial strains scientists had to produce new types of penicillin. In 1945 the chemical structure of penicillin was determined using X-ray crystallography, and it became possible to engineer new forms of penicillin. These succeeded, but eventually it became clear that the deadly microorganisms can never be totally eradicated, their threat can only be mitigated. Nevertheless, penicillin has been an extraordinarily successful drug; it has been estimated that it saved tens or even hundreds of millions of lives in the twentieth century, and continues to do so.

A variety of sophisticated medical imaging devices are now widely used. Electron microscopes can see viruses, just as traditional microscopes could see bacteria in the nineteenth century; they provide magnifications of up to ten million times, five thousand times greater than traditional microscopes. They work by illuminating a target with an electron beam and detecting the transmitted and scattered radiation in a variety of ways. X-ray imaging has been in use for most of the past century, following the discovery of X-rays by Wilhelm Röntgen in 1895. An X-ray beam can pass through the human body, but dense materials (such as bones) absorb the X-rays more than do other tissues, so they stand out clearly on the image obtained by a digital detector placed on the other side of the body. A computer tomography (CT or CAT) scanner uses a rotating frame containing an X-ray source on one side and a detector on the other; the frame rotates around the patient, so that cross-sectional images can be built up by a computer. These can then be combined to produce 3D images as seen from different angles.

Positron emission tomography (PET) employs a positron-emitting tracer that is introduced into the body; it emits very energetic photons (gamma-rays) that can be detected and again made into 3D images. Magnetic resonance imaging (MRI) is based on the absorption or emission of radio waves by certain atomic nuclei in the presence of an external magnetic field. It is a highly versatile imaging technique, and does not have the (now minimal) hazards of X-rays. Ultrasound is a very different kind of imaging technique. A rapidly alternating electric charge produces oscillations in a ceramic element through the piezoelectric effect (discovered in the nineteenth century), which in turn makes high-frequency sounds waves (far too high for us to hear). A pulsed beam of these waves is aimed at the target, and the reflected echoes are detected and displayed in two dimensions in real time. These are all impressively sophisticated techniques, and they certainly show that modern medicine makes heavy use of advanced scientific knowledge.

Sound totally mystified the early scientists. How could they 'hear' the visible movement of someone's lips, or a jar breaking on the floor? There appeared to be no visible causal connection between those events and the sounds they heard. The history of sound is rather murky. Aristotle was apparently one of the first to suggest that sound travels in waves, and Leonardo da Vinci has sometimes been credited with the same insight. Pythagoras found that the pitch of a sound is related to the length of a plucked string, and Galileo found more directly that it depends on the frequency of the waves. In 1640 Marin

Mersenne was able to measure the speed of sound, and some years later Otto von Gericke and Robert Boyle proved that air is the medium in which sound waves travel: placing a ringing bell into a vacuum jar containing no air silenced the bell. So it became established that sound waves are density waves propagating through the air. Over the last two centuries it has become possible to copy, manipulate and transmit sound in a wide variety of very clever ways.

Modern hearing aids are interesting examples of current technology. The standard miniature battery-powered hearing aid contains a tiny microphone that detects sound and converts it into an electrical current, an amplifier, and a loudspeaker to play the amplified sound via a small tube directly into the inner ear. In a digital hearing aid the amplifier chip digitizes the signals from the microphone, which can then be processed (as in a computer) to suit the environment and the specific needs of the user to give much clearer sounds. A somewhat different kind of 'hearing aid' is provided by noise-cancelling headphones, for use by anyone with normal hearing. They greatly reduce unwanted ambient noise (such as that in an airplane, or in the vicinity of leaf blowers), while at the same time playing high-quality music. The principle is simple: the ambient noise is measured by a microphone and a waveform is generated that is the exact opposite, to provide near-perfect cancellation in real time; meanwhile the user can enjoy unpolluted music or silence, as desired.

A cochlear implant is a very different and much more sophisticated kind of hearing device, providing a sense of sound to those who are profoundly deaf or severely hard-of-hearing. It bypasses damaged parts of the ear and directly stimulates the auditory nerve. It consists of an external part behind the ear and an implant that is surgically placed under the skin. In the external part the signals from the microphone go to an electronic sound processor, and the modified signals go to a transmitter. In the implant those signals are picked up through electromagnetic induction by a receiver which converts them into electrical impulses to be sent to an electrode array, which in turn stimulates the auditory nerve to send signals to the brain. Hundreds of thousands of these devices have now been implanted worldwide with high success rates, especially for young children. The U.S. National Institutes of Health considers the cochlear implant to be "one of the more ground-breaking biomedical achievements in the last 30 years". Interestingly, some of the strongest objections to the cochlear implant come from the Deaf Community, who regard it as an affront to their minority culture.

The heart is at the very core of our beings, and for millions of people over the past half century even that has been regulated by a product of science and technology: the artificial pacemaker. It is used to maintain an adequate heart rate when there are problems with the heart's natural pacemaker or electrical conduction system. The story goes back to 1889 in England, when John Macwilliam found that he could give electrical impulses to restore the normal heart rhythm in a patient. In 1926 Mark Lidwill and Edgar Booth in Sydney Australia made a device involving a needle plunged into a chamber of the heart, and used it to revive a stillborn infant. But public concerns were expressed about the idea of 'interfering with nature' in this way. Nevertheless, developments continued after World War II. The Canadian electrical engineer John Hopps made the first external (transcutaneous) pacemaker in 1950, and this was followed by a host of innovations. The first implantation of an artificial pacemaker was done in Sweden in 1958 by Rune Elmqvist and Ake Senning; the device initially failed a few times, but the recipient went on to receive 26 different pacemakers during his 86-year lifetime, and he outlived both the inventor and the surgeon. At the cutting edge today are pill-size wireless pacemakers that can be inserted via a leg catheter directly into the heart.

Even the creation of new human beings is now being influenced by science and technology. More than five million people alive today were 'created' by IVF (in-vitro fertilization), in which the egg is combined with sperm in a laboratory and later transferred to a uterus (of the mother or another woman). Soon one in ten British babies will be born via IVF. It really is quite remarkable that millions of people owe their very existence to science and technology. The first IVF baby was born in 1978, and the British physiologist Sir Robert Edwards was awarded the 2010 Nobel Prize in Physiology or Medicine for this achievement. At the other extreme, contraceptive pills were introduced in 1960. They are very effective in preventing ovulation, and reversibly inhibit female fertility; they are currently used by over 100 million women worldwide.

Regenerative stem cells are increasingly being used to cure or diminish various medical problems and diseases. Two breakthough studies have recently been announced. One of them appears to be a 'game changer' for multiple sclerosis (MS), in which the immune system attacks the nerves of the brain and spinal cord. For a group of patients with a type of relapsing remitting MS, chemotherapy was used to eliminate the defective immune cells, and these were then replaced by stem cells from the patient's blood and bone marrow. After 3 years this treatment succeeded in 94% of the cases, compared 40% for those given the standard treatment. The patients called it a life-changing 'miracle cure': after being in a wheelchair or unable to read they could once again live normal lives. The other breakthrough involves sight. Cataracts cause blindness, and cataract operations in which a patient's clouded lens is replaced by a clear plastic lens is the most common operation in the world. The new technique involves removing the clouded lens but leaving behind the lens capsule, a membrane which give the lens its required shape. Nearby regenerative stem cells are then inserted in the membrane and magically grow into a new, fully functioning 'living' clear lens. This has been successfully used for infants under the age of two that have congenital cataracts, and could potentially be used for the millions of older people with cataracts, restoring their sight using their own cells.

DNA is now being used in an increasing number of ways. Several fundamental scientific breakthroughs—the determination of the structure of DNA in 1953, the breaking of the genetic code in the early 1960s and the sequencing of the entire human genome in 2003—have opened up a wide variety of applications. One of them is DNA profiling (or 'fingerprinting'). 99.9% of human DNA is the same in every person, but there are small parts that differ, and these can be examined to distinguish between individuals, in some cases identifying a criminal or murderer. DNA testing can also identify genetic abnormalities, and it can provide information on family members and genealogy, including the major geographical locations of distant ancestors.

Genetic manipulation is now becoming commonplace. One of the first applications was in genetically modified (GM) crops, beginning in 1982. The objective is to introduce a desirable new trait into a plant which it would not otherwise have, such as resistance to insects and disease, and reduction of spoilage of crops. Methods have included 'gene guns' and microinjection to incorporate the foreign DNA (which has the desirable trait) into plant cells; it becomes integrated into the plant's own DNA. Recently more precise DNA editing techniques have become available, such as CRISPR. The concept has been widely accepted by farmers, and now over 10% of the world's arable land is planted with GM crops; in the U.S. over 90% of the planted area of cotton, corn and soybeans contains GM varieties. GM technology has resulted in much lower pesticide use and much higher yields. Scientists concur that GM crops pose no more of a health risk than conventional food, but many in the general public remain sceptical, and controversies continue.

Gene therapy is seen as a way of fixing a genetic problem at its source by modifying the genome, replacing or disrupting defective genes. T. Friedmann and R. Roblin introduced the concept in 1972 with their paper *Gene Therapy for Human Genetic Disease?* but urged caution in pursuing work of this kind. The first attempts date from the 1980s, but early failures dampened enthusiasm. Recently, however, there have been some notable achievements and growing promise for the future. In 2017 the U.S. Food and Drug Administration (FDA) approved gene therapies for two types of cancer. In both, the patient's own immune cells are genetically reprogrammed to target and kill the cancer cells. In the same year the FDA also endorsed the first gene therapy for a form of inherited blindness; this disease is caused by a single defective gene, and the cure involves a harmless virus genetically modified to carry a healthy version of the gene, delivered into the retina. In the U.K. another genetically engineered virus seems highly promising as a cure for haemophilia A (which prevents clotting in the event of severe bleeding). Recently scientists have directly edited the genomes of human embryos using CRISPR to remove and replace a single mutated gene segment that causes a common and deadly heart condition. The embryos in the experiments were not destined for implantation, and the research is a long way from clinical use, but will it eventually open the door to eradicating this and many other genetic disorders? And this is only the beginning; gene-edited pigs cleansed of harmful viruses may in the future provide transplants for human organs, and fertility-reducing genes may be implanted in undesirable insects—the possibilities seem to be endless, but so are the potential risks of unintended consequences— scientists will have to proceed cautiously.

There is an important distinction between two fundamental categories of cells: somatic cells are the differentiated cells of the body and tissues of an individual, and germ cells are those involved in reproduction. Gene therapy on somatic cells (SCGT) affects only the individual, but gene therapy on germ cells (GGT) is heritable and passed on to all later generations. There is strong opposition to GGT and in many countries it is prohibited (although 'medical tourism' may eventually become an industry). Aside from the obvious medical dangers of GGT there are strong ethical concerns, as it would raise the possibility (at least in principle) of 'designer babies' with certain 'preferred' traits—a new form of the much-loathed eugenics.

Materials science is now a burgeoning field at the interface of chemistry, physics and metallurgy. It covers materials from ceramics and polymers to metal alloys and semiconductors, and is fundamental to many industries. Graphene is a remarkable example. In 2004 Andre Geim and Konstantin Novoselov produced single-atom-thick sheets of carbon. Their paper was twice rejected by *Nature* (as being impossible), and finally published in *Science*. They were awarded the 2010 Nobel Prize in Physics for this fundamental development. Graphene is the first two-dimensional material ever known, and it is the lightest, thinnest and strongest material ever tested (200 times stronger than steel). It conducts heat and electricity (far better than copper) and is nearly transparent. There is a huge number of potential applications, including flexible semi-transparent mobile phones and lightweight airplanes. Nanotechnology goes even deeper than 'conventional' materials science, and involves the manipulation of matter on an atomic or molecular scale, already resulting in thousands of applications (including 'gecko tape').

When we think of the 1960s Apollo manned missions to the Moon, we tend to think of the complex orbital dynamics (based on Newton's laws), the sophisticated computers (based on quantum mechanics) and the rocket propulsion systems (based on Newton's third law). But equally important, if relatively unspectacular, was the materials science that went into the fabrication of the huge propellant tanks; their walls had to be strong but extremely lightweight (therefore thin), and they could not have been made with the technologies of the previous decades. The Apollo program involved a huge number of scientific and technological advances.

Meteorology is one of the atmospheric sciences and provides our weather forecasts. It was only when atmospheric physics was well understood and used in large computer simulations that major improvements in weather prediction became possible over the last several decades. Inputs for the models are provided by weather stations on land and weather buoys at sea, radiosondes launched into the troposphere and stratosphere, and weather satellites, and the models are updated every 10–20 min. Humans are still required to monitor the modelling and interpret the results, but human input may no longer be needed at some point in the future.

The above examples only scratch the surface. There are countless other applications of science—in our homes, offices, factories, roads, bridges, automobiles, trucks, trains and airplanes—in all aspects of our lives. Having scoured the natural world for knowledge of its contents and how it all works, we have been able to use that knowledge in a huge variety of applications for our own benefit, through ingenuity, creativity and industry. And both our scientific knowledge and the technology it spawns continue to grow at an exponential rate. We have been able to control and adapt many aspects of the world around us to greatly improve our lives.

What is really striking is that most of the technical developments that have made our world 'modern' have taken place just since the late 1800s—less than two lifetimes—nothing compared to the thousands of years of recorded history and the millions of years of human evolution.

In fact, if one were to pinpoint the single century that experienced the most dramatic developments in *all* of history, it would probably be 1870–1970. That 100-year period saw the discovery of the electron (basic to all electronics), the introduction of Einstein's relativity, the discovery of the universe and its expansion, the development of quantum physics and the discovery of DNA and the genetic code (the basis of all life), and we produced the 'horseless buggy' (the automobile), the telephone, the radio, the airplane (the one-time-only transition from not flying to flying), the electric clothes washer (which helped to emancipate women), the record player (those who had never heard a symphony could now hear one at the flick of a switch), movies, television, the atomic bomb and atomic power, the computer, jet airplanes, all culminating with the spectacular landing on the Moon in 1969—an immense achievement (and all this in spite of two World Wars and a Great Depression). Developments since 1970 have not been so obvious and dramatic, except for the monumental Internet and communications explosion, but science and technology have certainly been advancing in more subtle ways at a breakneck pace right up to the present.

One last comment on science and technology. Both are huge topics, but it seems far easier to review science than technology because there is just one reality, while there is an enormous number of diverse technologies that can be developed from it.

4.7 Science and Society

The above examples show how much science is involved in countless things that are part of our everyday world. Science has transformed our lives, and is deeply engrained in our modern civilization. It is probably impossible to estimate the economic value that science has added to society, but it is obviously enormous. Science is knowledge, and knowledge is power (*scientia potentia est*). But even aside from its great utility, science has been a huge intellectual triumph—probably the greatest achievement of humanity.

The very principles of science have become common currency. The idea that something has to be tested before it can be accepted seems obvious to us. 'Evidence-based knowledge' is the gold standard. Even television advertisements proclaim that their products "have been scientifically proven"; the word 'scientifically' is clearly meant to be the ultimate stamp of approval. Science is one of the most respected professions in the world.

But most people would be surprised to hear that it has been *curiosity-driven* research that has given by far the most profound, wide-ranging and important contributions: Newton's laws of physics, Faraday and Maxwell's work on electromagnetism, Darwin's theory of evolution, Einstein's theories of relativity, quantum mechanics which underpins all of chemistry and electronics, and Crick and Watson's discovery of the structure of DNA (the basis of life), to name but a few—all done out of pure curiosity. Curiosity-driven research has made our modern world possible, but even today it is harder to obtain funding for 'frivolous' curiosity-driven research than for 'serious' goal-oriented research.

Public attitudes to science have varied over the years. While no one today would doubt the power of science, there have been and still are concerns about some of the creations of science, as discussed in the next section. It is a very positive thing that members of the public are aware of scientific activities and speak out when they see a potential problem. Scientific research—aside from that done for military purposes—is open to public scrutiny. And the scientists themselves openly criticize each other when appropriate. If it is felt that one or more scientists are moving in a direction that may have negative consequences, others will call attention to it. This openness is one of the strengths of science.

Public interest in science is very important. The world of science can be especially appealing to young children, stirring their interest and imagination. They are excited by the planets in astronomy, the volcanoes in geography, and the insects in biology, and this may lead them into careers in science. Adults too can find new scientific and technological developments amazing and exciting; their appreciation of the fact that science has given us both our cherished knowledge of the world and our high standard of living today is what leads them to pay for the scientific enterprise through the public purse.

'Citizen science' is on the rise, thanks to the wide availability of computers, smartphones and sensors of all kinds. An early version was in the form of volunteering the spare processing power on home computers linked up for professionals to more efficiently do calculations or hunt for alien signals. But citizen scientists are now more active participants, examining images and huge databases for exceptional phenomena, looking for celestial events, helping to fold model proteins and DNA online, monitoring atmospheric conditions, air quality, soil conditions or water temperatures in out-of-the way places, making wildlife surveys, and recording the songs, communication and behaviour of animals, to mention but a few. It is a win-win collaboration, enabling citizens to enjoy the science and the sense of contributing, and expanding the scope of the scientists' work.

For the most part science advances without causing much of a ripple. What were the major events that have exploded into public awareness since the beginning of the Scientific Revolution? Galileo's small book *The Starry Messenger*, containing his astronomical discoveries, would certainly count as one; it rapidly became famous throughout Europe. Newton's *Principia*, which changed the world, was another. Darwin's masterpiece *On the Origin of Species* was so compelling that it was widely accepted almost immediately. Einstein's theory of special relativity was seen as a revolution in science, and the detection of the bending of light by the Sun, which he predicted, was a spectacular event. Then of course the atomic bomb, which exploded onto the world stage. The determination of the structure of DNA in 1953 caused great excitement. The creation of artificial life, if and when it happens, would be monumental. And the discovery of extraterrestrial life, most likely of microbial form on Mars, would change our view of our place in the universe forever.

But, for all that, science has a communication problem. The modern scientific and technological world we live in is both daunting and baffling to many. Granted that wonders such as electricity, computers and airplanes are undeniably real and indisputable reminders of the power of science, there are still quandaries and puzzles in our everyday lives and our views of the world. Is the fluoride in our water really safe? Should our children have vaccinations? Are we really descendants of the apes? We have common-sense views about the world, and science often seems counter-intuitive. Then how do we make decisions in our lives? We try to be rational, but it is hard to suppress our natural beliefs and those of our community. There is an understandable tendency to rely more on the opinions of friends and family than on experts, abstract statistics and the cold technical jargon of science and technology. We want absolute certainty, and we are unsettled to learn that scientific knowledge is always provisional. But scientists also have to deal with these issues. They have to submit their results for formal peer review before they can be published. Through this process they are made well aware of the high standards set for the vast body of rigorously established scientific knowledge. It is the best that humans can do. Science provides both the most reliable knowledge available and also a way—the scientific method—of making decisions based on evidence. Many scientists make great efforts to explain the wonders and powers of science to the general public, but there is still much to do.

In the midst of all this there is a thriving cottage industry of doubting. The Flat Earth Society still exists. It has long been claimed that the Apollo Moon landings were faked. The fluoridation of water, widely accepted half a century ago, still incites angry demonstrations in some communities. The anti-vaccination movement by some seriously undermines efforts to immunize the wider community. Opposition to genetically-modified (GM) crops continues even though GM is no more dangerous than traditional breeding. And of course there is visceral opposition to the notion that anthropogenic climate change is taking place. Doubters delight in fighting the consensus of experts. Conspiracy theories abound and will never die. What is more sinister is that scepticism is sometimes promoted by vested interests, such as the fossil fuel industry in the case of climate change; all they have to do is sow the seeds of doubt. Years ago, edited and refereed books, scientific journals and encyclopaedias provided the highest-quality access to scientific knowledge. But in today's world there is fake news, fake science, and fact-free 'bubbles'. And the Internet makes it easier than ever for science doubters, who can live in their bubbles amplifying their message along with similarly-minded doubters around the world. These are new problems for society.

In the face of these issues science can only do what it has always done—provide the best possible evidence-based knowledge about the natural and physical world we live in. It is essential that the high professional standards be maintained, so that the body of scientific knowledge is always a reliable, neutral and unbiased resource that is equally accessible to everyone.

Scientists are recognized for their contributions, and various prestigious awards are given each year for ground-breaking discoveries and developments in science. The Nobel prizes are by far the most famous, and they endow the recipients with almost celebrity status. They were established in 1895 by the Swede Alfred Nobel, who made his fortune by inventing dynamite. He allocated 94% of his assets to establish three Nobel Prizes in science (Physics, Chemistry and Physiology or Medicine) and two others (Literature and Peace). The total assets of the Nobel Foundation currently amount to about $520 million, and each prize (often split between two or three recipients) in 2017 was $1.1 million. Each year the three science prizes are awarded 'for services to humanity' by Swedish academies, having considered nominations from thousands of scientists around the world, and the magnificent prize ceremony takes place in Stockholm, the prizes being given by the King of Sweden.

While the Nobel prizes have been highly beneficial in honouring and promoting science, there have of course been controversies. Over the 116 years since the first Nobel Prizes were awarded, women have received only 2%, 3%, and 6% of the prizes in physics, chemistry and medicine respectively. No black person has yet received a Nobel prize in science. The Nobel Prizes often consider entire careers as well as the prize-winning discoveries. Since 1974 they have not been awarded posthumously. There is often a long time-lag (even decades) between the discovery and the award, which now typically comes much later in life. As no more than three scientists can share a single prize, it can seem a bit unfair to the others in large modern collaborations involving hundreds or thousands. Even in recent times (2010–2017), over two-thirds of the Nobel prizes in science still go to individuals born in Europe (39%) and the U.S. (30%); Japan has risen to 15%, but China and the other emerging Asian giants have not yet made a significant mark. Several prizes were highly controversial. And the Nobel Prizes skew the view of science: they omit a number of other front-line sciences.

Today other similarly generous prizes are awarded in some of these other fields. In 1980 the Crafoord Prize was established by the Swedish industrialist Holger Crafoord and his wife for disciplines not covered by the Nobel Prizes. This award, presently amounting to $750,000, is given annually on a rotating basis for Astronomy and Mathematics, Geoscience, and Bioscience. As in the case of the Nobel Prize, the Royal Swedish Academy of Sciences selects the laureate, and the prize is presented by the King of Sweden (in a far more intimate ceremony). Recently some mega-prizes have come onto the scene. In 2012 Yuri Milner and other entrepreneurs introduced the Breakthrough set of prizes for the life sciences, fundamental physics and mathematics; at $3 million these are the most lucrative academic prizes in the world. Property developer Samuel Tang introduced the Tang Prize as an Asian complement to the Nobel Prize. There has been some controversy over the motivations for these prizes, the large sums involved, and the distortions they may create in pure and untainted science.

* * *

Science and technology have dramatically changed the way we live and work. The Industrial Revolution brought masses of people from the countryside to the factories in towns and cities. There was much hand-wringing over the dislocations this would cause in society, but the world adapted. In 1700 only 19% of the British lived in towns and cities; today 90% do. In 1800 only 5% of Americans lived in towns and cities; today 82% do. In 1850 67% of all Americans lived on farms in the countryside; today only 2% do. In 1960 24% of American workers were in manufacturing; today just 8% are. First there was a rush from the countryside to the jobs in factories. Then robots took away jobs in the factories. And recently computers have been taking away jobs in the office. Still the world has adapted. These are sometimes called 'disruptive technologies'. An extreme recent case was photography: in just a decade it changed almost completely from film to digital, and Kodak, with over 100,000 employees at its peak, went bankrupt.

Objections to machines taking jobs away from workers go back hundreds of years. In the 1580s the hosiers' guilds in Britain objected to the introduction of a knitting machine that would make their skills redundant. In the late 1700s and early 1800s the 'Luddite movement' violently opposed the use of automated textile equipment and destroyed machines, resulting in the death penalty being imposed for 'machine breaking'. In the 1930s the famous economist John Maynard Keynes predicted that the 15-h work week would be the norm for his grandchildren (if anything, we're now working 15-h *days*). As wave after wave of new technologies were introduced over the years, the net results have been and still are improved standards of living, wage increases in line with productivity, workers adapting to new occupations, and low to moderate unemployment.

Can this go on forever, or will there one day be 'a world without work'? One of the reasons (along with globalization) given for the recent populist movements shifting national politics in some countries is the disconcerting pace of relentless technological change and the sense felt by some members of the population of being 'left behind'. In response to all this there is increasingly talk of a 'Universal Basic Income' (UBI), to be paid to all citizens of a country to provide for the chronically unemployed. The idea has been around for a long time, but it is gaining traction and several countries are now running trials. Switzerland actually held a referendum in June 2016 on introducing a UBI, but it was rejected by 77% of the voters. At that time the unemployment rate in Switzerland was only 5.1%, and is currently 7.3 % in the EU, 3.8% in the U.S. and just 2.5% in highly-robotized Japan, so the idea of a UBI seems premature; at least on a national scale there is no indication yet of that new world without work.

But—is this time different? Are we at an inflection point? Will robots and 'artificial intelligence' (AI) soon be able to do almost everything? The unprecedented exponential growth in science and technology in recent years has led to robots and computers so powerful that they may now be able to take over many jobs that we previously thought only humans could do. Some very sophisticated studies have recently shown that up to half of all jobs in advanced countries may be at high risk of being taken over by automation within just the next 10–20 years; one study estimates that automation could affect as many as 800 million jobs worldwide by 2030. These are mostly jobs that are repetitive and routine, and include occupations in transportation and logistics, office and administration support, sales and services, and labour in production. Telemarketers, tax preparers, legal assistants, fast-food cooks, bank tellers, data entry keyers, file clerks, loan officers, labourers and a host of others are at high risk. The workers in many of these jobs tend to have the lowest education and the lowest wages. The occupations with the lowest risk of automation are those that require significant cognitive, creative and social skills, and these are correlated with high education and high salaries. Doctors, dentists, nurses, healthcare workers, teachers, lawyers, engineers, scientists, architects, and human resource and senior managers are in this category.

However, while automation can displace labour it can also complement it and produce entirely new industries, resulting in productivity increases and employment growth. And, as often before, the displaced workers should hopefully be willing and able to move and find alternative employment in these new industries and others. The ground is shifting and workers have to be adaptable. The days when a worker has only one job and one employer in a lifetime are fast disappearing, and in the future a career may involve a series of different jobs. The big danger is that (at some time in the future) many of the workers displaced from their jobs may no longer be able to find work that they can do but computers and robots cannot—technology is moving up the food chain. They would be chronically unemployed. Can this be avoided forever? If not, governments would finally have to resort to reducing the workweek or ultimately introducing something like a UBI. It has been suggested that a UBI could be paid for by a 'robot tax' to redistribute wealth from the robot owners to the chronically unemployed. But unemployment is a social problem, not just a financial one. What will that future world look like? We can only hope that, as in the past so in the future, society will adapt.

Education is the way to keep ahead of the curve, and in the globalized marketplace with offshoring of jobs this means a race between the countries of the world. The OECD (Organization for Economic Co-operation and Development) keeps track of the relative quality of educational systems in 72 countries of the world in its triennial survey called PISA (Programme for International Student Assessment). Over half a million students, representing 28 million 15-year olds in the 72 countries, took the two-hour test in 2015; they were assessed in science, mathematics, reading, collaborative problem solving and financial literacy. The top ten countries in science were (in order) Singapore, Japan, Estonia, Chinese Taipei, Finland, Macao (China), Canada, Viet Nam, Hong Kong (China) and B-S-J-G (China).[4] Britain, Germany and The Netherlands were tied in 15th place, and the U.S. was in 25th place. Singapore was again first in reading, followed by Canada and Hong Kong. In mathematics the top three were Singapore, Hong Kong and Macao.

Science education is essential not only for budding scientists. Science is a major part of our heritage and culture, and it is important for citizens, leaders and decision-makers to have a good understanding of the broad concepts of science and its role in underpinning modern technology and society. Science is at the heart of our knowledge-based economy. But an education in science is more than just learning 'facts'. It is a *way* of knowing; it teaches us a powerful way of thinking: making hypotheses and experiments, collecting data and evidence and using logical reasoning to arrive at understanding and knowledge. It teaches us how to solve problems and analyse and discuss matters of importance to all of us, in all fields. Science should be a part of everyone's education.

But the traditional ways of teaching science had their shortcomings. They were focused on the basic facts, laws and theories rather than the broader

[4] Beijing-Shanghai-Jiangsu-Guangdong.

concepts of science and the dynamic and evolving nature of scientific 'truths'. They emphasized memorizing rather than thinking. They could be dull. It was easier to test for scientific words or facts than for scientific understanding. This was highlighted by the use of multiple-choice questions for the 'right' answer; it trivialized science education. The old system rewarded those who might continue to study science at higher levels, rather than encouraging every student to be engaged with science. This resulted in declining interest in science and science careers by many students around the world. School science was seen as abstract and irrelevant. There was little encouragement for ideas and free discussion. Science was also seen as difficult (especially when combined with mathematics)—why study science when there are more interactive, interesting and less difficult subjects? The stereotypes of scientists didn't help: old men in white coats working with test tubes in the stuffy and dark recesses of laboratories. And there was public and parental apathy towards science.

This situation and possible remedies have been discussed in several documents in recent years. Among them are the U.S. National Research Council of the National Academies report *Taking Science to School* (2007), the UNESCO study *Current Challenges in Basic Science Education* (2009), and the OECD's *PISA 2015 Results in Focus*. The brief paragraphs here summarize some of the main points.

All of these documents stress the need for a more dynamic approach to teaching science. Students should be actively involved, not just passive listeners. The old ways concentrated on transmitting information, but the modern approach includes discussing and understanding concepts. Today we can find the facts on the Internet, but we need to know how to use them. Students should know how science is really done, including remarkable stories of intuition, creativity, discovery and serendipity. When students take part in experiments, they can see science happening before their eyes. They can feel the excitement, even passion. In critically discussing and interpreting the results they are working like scientists. This gives them motivation and self-confidence. And working together in groups, making decisions and rational conclusions they are developing skills that can be applied in many areas of adult life.

Teachers are the key players. How they teach, lead and inspire are strongly correlated with the performance of their students. Finland is everyone's favourite educational success story. There, teaching is the most prestigious occupation. Only one in ten applicants make it into one of Finland's eight teacher-training schools, and almost all of them stay in the profession until retirement. Finnish teachers' salaries are high relative to similarly educated workers. And the results show it: Finland ranks fifth worldwide in the PISA scores for science. Singapore, which is top in all three of the PISA categories, rotates groups of outstanding teachers through the highest levels of the education policymaking apparatus. Continuing education is important. The teachers get the respect they deserve, and this stimulates them further in their work with students.

Because it develops cognitive, assessment and communication skills in addition to teaching the fundamental concepts of science, science education can be as important for students who go into other occupations as for those who go into science careers. In China there is a move to include science as a core subject in basic education from kindergarten on up.

Now more than ever there is the possibility of outside support for science education. Locally there are science clubs and associations, science museums, visits from professional scientists, science competitions and science fairs. But the biggest developments come from the Internet. There is a huge range of material freely available on the Internet, including experiments performed by and lectures given by some of the best experts in the world, websites of science museums around the world, global collaborations with other schools and science associations, and global science competitions. Collaborations can take place between teachers, researchers and curriculum developers. It's a whole new world for science education.

But in this world of over 7.6 billion people, not everyone has the opportunity to have a career in science. Far from it. The world is losing out on the potential of tens of millions of individuals who could excel at science if they only had a chance. Even today the advanced societies do most of the science, and most of the scientists are male.

Many scientists in the past have indeed risen from humble beginnings, as is clear from the brief history in Chap. 2, but these were mostly young men living in the Western world. The rest of the world is now struggling to catch up, but there are many obstacles.

In the Third World the hurdles are immense: poverty, prejudice, culture, language, poor schooling, and an almost total lack of opportunity. In the small villages in many of these countries the family and village life are everything, and the outside world is relatively unknown. Education is limited, expectations are low, and the immediate needs of the family take priority. Even leaving the village to attend an institution of higher education in a local town can be a jolting experience. Efforts are certainly being made by governments and aid organizations to improve the situation, but progress is slow.

Most countries have at least a few universities, and the governments make some efforts to make university education as accessible as possible with subsidies and loans. But in many countries there are few opportunities for a career in science, and the financially more rewarding prospects of business and banking tempt many to go in those directions. So many of those who want to persevere with science have to make the hard choice to leave their families and country and go abroad for advanced training and careers.

India has a large number of universities, and has had some top-class research institutes for decades. But it also has a huge population (1.3 billion), many still living in small villages, and the hurdles mentioned above apply here also. It has well over a thousand languages and dialects, of which 22 are 'official'; Hindi is by far the largest, spoken by over 40% of the population, and—a huge advantage—India is the second-largest English-speaking country in the world, with about 125 million English speakers. Unfortunately, there is still significant discrimination based on the caste system of Hindu society, although this is constitutionally illegal, and the government has provided quotas and other affirmative actions to support the lower castes. It was thought for some time that the very existence of many centres of scientific excellence would eventually 'trickle down' to the masses, but it has been realized that a more pro-active 'bottom-up' approach may be better in finding talented young people in the poorest villages.

China currently produces more science PhDs than any other country in the world. The communist government is trying to make education equally accessible to all. To provide for all of its 55 ethnic minorities, it has established bonuses and quotas. University tuition fees are low, and there are scholarships and loans. But inevitably children of government officials and wealthy businessmen have an advantage. Salaries for post-docs and scientists are relatively low, and many of the best are lured to greater opportunities in wealthier countries.

Even in wealthy countries, many children from underprivileged backgrounds struggle with poor education, dysfunctional families, discouragement from their neighbourhoods and peers regarding intellectual achievement, and inadequate funds to attend universities. Some of these problems are frequently associated with race. And science courses can be among the most expensive to attend. Top universities in the U.S. can cost $60,000 per year. There are various sources of financial assistance—parents, scholarships, grants—but often a student does part-time work, which detracts from achievements at university. But there is some relief at the top end—in postgraduate programmes tuition is often covered by grants and teaching positions.

Half of the world's population, 3.8 billion people, are females, but they have been grossly under-represented in science over the years. In many poor and developing countries, girls and women have been restricted, both from education and from careers. And in many of the 'advanced societies' the governments did not allow married women to work (except during the wars) until the second half of the twentieth century, and even single women were largely excluded from the major professions. And as of November 2016, only 49 of the grand total of 911 Nobel prizes were awarded to women.

But this has been changing dramatically over the last several decades. Now upwards of half of the students in science and technology in the advanced countries are women, and in 2011 women accounted for 20, 30, 40 and 50% of the PhDs in physics, mathematics, chemistry and the biological sciences respectively in the U.S. Worldwide, women now represent half of the graduates at the bachelor's and master's levels. Women have held the top positions in major scientific organizations such as the European Organization for Nuclear Research (CERN) and the European Southern Observatory (ESO), and in major technical corporations such as IBM and Hewlett Packard. And of course women have been leaders of several countries including Argentina, Australia, Brazil, Canada, Chile, Croatia, Finland, Germany, Iceland, India, Indonesia, Ireland, Liberia, Lithuania, Nepal, New Zealand, Nicaragua, Pakistan, Philippines, South Korea, Sri Lanka, Switzerland, Taiwan and the United Kingdom. So the glass ceiling has certainly started to shatter, but there's still much more to be done, especially in the underdeveloped countries.

4.8 The Powers and Perils of Science

By innocently scouring the world for the secrets of nature out of pure curiosity, scientists may inadvertently make possible some potentially perilous developments. This is sometimes unforeseeable. This happened, for example, once we were able to reach inside the nucleus of an atom, and inside the nucleus of a living cell.

When we reached inside the atomic nucleus we discovered that energy can be extracted from it. In the 1920s the scientists studying the atomic nucleus were doing so out of pure curiosity and in great excitement. There was no concern about what this research could lead to. In the 1930s some scientists were becoming aware that energy could be extracted from the nucleus of an atom, but initially even Einstein was doubtful that it would ever become a reality. The idea of a nuclear chain reaction was first proposed by the Hungarian Leo Szilard in 1933. In 1938 the German chemists Otto Hahn and Fritz Strassmann discovered nuclear fission; the theoretical explanation was soon provided by Lise Meitner and Otto Frisch. Enrico Fermi and colleagues produced the first nuclear fission in the U.S. in 1939, and Szilard and Walter Zinn proved that a chain reaction was conceivable. As most of the scientists in this field had lived and worked in Germany, there was a fear that the Nazis might develop an 'atomic bomb' based on these developments. Szilard drafted a letter to President Roosevelt warning him of this possibility and suggesting that the U.S. should start its own nuclear programme; Einstein was persuaded to sign the letter in August 1939, adding his immense prestige to the proposal. Thus was born the Manhattan Project.

Germany never did develop such a bomb, but the Manhattan Project succeeded; the first atomic bomb was exploded in New Mexico in July 1945, and in the following month two bombs were dropped without specific warning on Hiroshima and Nagasaki in Japan, ending the Second World War.

Many books have been written about the Manhattan project, and the views of the scientists involved. There had been a sense of exuberance in working on the project, but when word came back of the terrible destruction many of the scientists became introspective and profoundly disturbed, and many felt guilty. In October 1945 J. Robert Oppenheimer, one of the most prominent 'fathers' of the atomic bomb, visited President Harry Truman to persuade him to support international control over nuclear weapons. But some other scientists went on to develop the much more powerful hydrogen bomb.[5]

The concerns of society reached a peak in the 1950s with the 'ban the bomb' campaigns. Nevertheless, even today, decades after the Cold War, we are still living in the shadows of the world's nuclear arsenals, and a small misstep could quickly lead to nuclear oblivion.

The atomic bomb certainly made everyone aware of the power of science, but also deeply concerned, not only about the bomb, but also about what the future of science might bring. Should everyone stop doing science? But if it doesn't continue in one place, it will undoubtedly continue in others. The increasing number of countries that have developed the bomb is a constant reminder of this problem.

When we reached inside the nucleus of a living cell we discovered the very basis of life, DNA. While this was an exciting and fundamental discovery for science, it also opened the way to a host of potential manipulations and applications. So genetics is another field of science that has provoked concern over the years, although public attitudes vary depending on the specific

[5] The atomic bomb is based on the splitting of uranium or plutonium nuclei, and the hydrogen bomb is based on the fusion of hydrogen nuclei to release energy. As physicists probed deeper and deeper into fundamental physics, would they find more subatomic processes that release huge amounts of energy? After almost 70 years of further research, the answer is yes. In late 2017 Marek Karliner and Jonathan Rosner showed that quarks—the constituents of protons and neutrons—can also fuse and release an amount of energy similar to that of hydrogen fusion in one case and ten times greater in another. At present the very short lifetimes of the quarks preclude any practical applications.

application and technology. Genetic testing is viewed positively and cloning is viewed negatively (we have so far cloned 23 mammalian species, including our close relative the chimpanzee). Attitudes tend to be positive for uses that maintain the natural order and negative for uses that change it. The development of genetic research has been gradual, and there has been no sudden shock that has caused a crisis in public opinion. The beneficial applications of genetic science are growing, and pointing to a revolution in medicine over the coming decades.

But there have been significant controversies over some developments, in particular genetic engineering, changing the very DNA of an organism. While it is true that the selective breeding of plants and animals (carried out for thousands of years) results in modifications of the genome and is therefore a kind of genetic engineering, it was never controversial. But the more recent innovations have caused a stir. As mentioned above, genetically modified (GM) crops are now used quite widely and give several advantages. Many farmers see them as beneficial and scientists concur that they do not pose a health threat, but many in the general public remain sceptical. Gene therapy, also mentioned above, involves direct meddling with the human genome, and, in view of the immense complexity of both the genome and the cell that contains it, many scientists as well as the general public are concerned about the likelihood of unintended consequences. There are also concerns that gene therapy may be used not just for treatment of diseases but also for modification or enhancement of various human traits such as appearance, physique, intelligence or character. Leading scientists in major world organizations have called for a moratorium on *heritable* human genome changes, and many countries have flatly prohibited any such gene therapy. It will probably be a long time before these views change, but in the meantime there always remains the possibility of this kind of research continuing in other countries.

The controversies are wide-ranging: religious accusations that the scientists are "playing God" issues of the "patenting of life" and in the specific case of GM crops criticisms of the regulatory processes, the labelling of products, worries about gene flow and contamination of other organisms, and environmental and health issues in general. There is an underlying concern about meddling with the natural way of things and upsetting the delicate balance of nature in general.

For more than a decade a controversy has raged over the use of pluripotent embryonic stem cells. These are undifferentiated cells that can become any type of cell, and can propagate indefinitely. They have many advantages over the 'adult' stem cells previously used, but obtaining them results in the destruction of the embryo. This caused a storm of protest based on ethics. Is it justified to destroy an embryo in order to cure many patients? At what point does life begin? Several countries have banned the use of embryonic stem cells, and it remains controversial in others. Recent discoveries may eliminate the need for embryonic stem cells, and this new research may even open up a whole new possibility: reversing the effects of ageing in at least some cells.

Artificial life is another highly controversial field. Various steps in that direction have already been taken, producing self-replicating organisms with artificial DNA and even a completely new genetic code with six letters rather than the usual four. While this kind of research is exciting and may someday show how life on Earth could have formed from abiotic molecules, it is at the same time worrying; if true artificial life were ever to be created, it would certainly cause a great shock, as the ramifications would be profound. And, of course, it is reminiscent of the ghoulish science fiction novel written by Mary Shelley in 1818 about Frankenstein, a scientist who created a grotesque but sapient creature.

As more and more labs around the world are able to do relatively sophisticated work and more powerful tools such as CRISPR become available, the possibility of a killer virus and pandemic being produced inadvertently or on purpose always looms on the horizon. And 'do-it-yourself biology' is now possible for amateurs, using many of the same techniques as the mainstream research institutions. It is the era of 'garage biology'. Is the genie being let out of the bottle? Efforts have been made to regulate this activity, but so far they have been relatively ineffective, and the self-described 'biohackers' feel that openness and transparency are better than regulations.

Biological and chemical weapons pose a continuing threat to humanity. Biological weapons are based on disease-producing agents such as bacteria, viruses, fungi and toxins. Their use goes back for centuries: plague-infested bodies thrown over fortified walls during battle or smallpox-infested blankets given to indigenous populations, for example. The 1972 Biological Weapons Convention prohibits the 180 signatories from the development, production and stockpiling of biological and toxic weapons, but there are continuing concerns. While smallpox has been officially eradicated from the world (a huge victory for public health in 1980), approved samples are preserved in two labs, one in the U.S. and one in Russia, and it is always possible that the virus may fall into the wrong hands. As virtually the entire population of the world now has no immunity to smallpox, a tiny release of a fine spray (for example in a major airport) could spread the disease rapidly over the globe. And there are many other potential (and possibly synthetic) biological weapons; as information on their manufacture is widely available on the Internet, their use by a rogue state or terrorist organization is a constant danger.

Chemical weapons are distinct from biological weapons; they use toxic chemicals that can inflict harm or death. Examples are mustard gas, phosphine gas and various nerve agents. Some were discovered accidently decades ago by researchers trying to develop pesticides. Although the U.N.-sponsored Chemical Weapons Convention of 1997 was signed by 192 countries and over 93% of the world's stockpile has been destroyed, there remain concerns; stockpiles are maintained in some countries as a precaution against any aggressor. It has been argued that wholesale biological or chemical warfare is unlikely because it could backfire or even cause a global pandemic; however it seems unlikely that we will ever be completely free of the potential threat of biological and chemical weapons.

One would not think that Darwin's theory of evolution by natural selection would be one of the perilous scientific developments. But unfortunately his theory encouraged his cousin Sir Francis Galton to conceive the idea of what he called *eugenics* in 1883 to improve the human population (although Darwin himself strongly disagreed with the idea). Desirable human qualities (presumed genetic) were to be encouraged, and undesirable qualities discouraged. The idea gained traction and was in full swing in the early twentieth century in many countries. Altogether some 60,000 people were eventually sterilized in the U.S. alone; Nazi Germany, wishing to establish a 'pure' Aryan race, took eugenics to well-known extremes, showing how terribly wrong things can go. Eugenics was reviled and collapsed after World War II. Fortunately, modern genetic analysis is breaking down ancient racial stereotypes by revealing the complex interactions that have taken place between societies in the past.

The power of modern electronics has led to a rapid decrease in privacy. The huge computers and data banks of the world contain information on all of us. Banks have electronic records of our savings accounts, investments, credit card numbers and financial transactions. Insurance companies know our medical histories, home and car details. Everything voluntarily put on a Facebook page remains forever—a host of information on details of our lives, friends, activities, political views, worries and aspirations. Web browsing leaves an electronic trail, and on-line shopping gives information on our personal preferences and more. Our smartphones always know where we are, and large numbers of security cameras can keep track of us using face-recognition techniques. Doctors and hospitals increasingly share our medical details. Governments are collecting security information including fingerprints and photos as well as social security numbers, passport numbers and our traveling patterns as we pass through airports.

We quietly acquiesce, as we know these things are beneficial to us, but they also bring new hazards. Information can get into the wrong hands. The data on hundreds of millions of people can be suddenly stolen from a major bank. 'Identity theft' can cause major problems for the victims. And some government agencies are quietly trying to amass as much information on their citizens as they can, vaguely reminiscent of 'Big Brother' in George Orwell's book *Nineteen Eighty-four*. It may seem far-fetched to be concerned about this, but there have been times in the past when government records were used for evil purposes, such as the well-organized Dutch municipal records that made it easy for the Nazis to find Jews. The loss of privacy in the modern world is unprecedented. Can we live with it?

Is artificial intelligence (AI) a threat to us? Is there a danger of AI 'going rogue' and spelling doom for humanity? Can we even now not understand the decisions made by some AI machines? There has been some hysteria about this possibility in recent years, fuelled by decades-old science fiction and the now rapidly growing power and potential of cutting-edge self-learning AI systems and robotics. Certainly AI is a major factor in automation replacing jobs as discussed in the last section and in the loss of privacy as mentioned above. But AI robots are light-years away from being anything remotely like the conscious and sentient beings of science fiction (and some would argue that such developments are completely impossible, even in the far future), so the hysteria seems unwarranted. Furthermore, they are after all just human creations (even if they are capable of self-learning), and humanity can always keep its finger on the 'abort' button. Nevertheless, it seems prudent to establish a regulatory environment to monitor current developments, which involve billions of dollars per year, and assure that AI always does just what we want it to do and no more. So far AI is our friend and servant, in the computers and smartphones we use, in search engines, face and speech recognition, and soon in self-driving cars and medical diagnoses.

It is said that the computer-human combination is far more powerful than computers alone or humans alone—a winning partnership in games like chess with the computer racing through all possible moves at incredible speed and the human adding intuition and judgement. And the marriage of AI with neuroscience is already starting to show results. With a brain-computer interface involving a computer communicating with a chip embedded in the brain, a physically handicapped person can move a computer cursor, control a motorized wheelchair and move a robotic arm by pure 'thought' alone. And researchers can already distinguish between simple 'thoughts'—between thinking about tennis or bed, for example—using functional magnetic resonance imaging. The day may come when computers can more generally 'read' mental processes and obey the instructions of the brain, transforming the lives of seriously handicapped people and enhancing the lives of others. Large investments, both public and private, are being made on this new frontier. But again there are serious concerns. To what extent will all this invade private lives and change the very concept of what it means to be human? Where do we draw the line? There are now calls for this kind of research to be regulated, but it is advancing in so many ways by so many researchers in so many countries that it may not be easy to contain.

There is certainly one arena in which AI and robots will not be benign. 'Killer robots' may bring the third revolution in warfare. Autonomous weapons systems could enable armed conflict on a scale far greater than ever before, on timescales faster than humans can imagine. They could be used by terrorists as well as nations. Autonomous weapons systems are on the cusp of development right now; more than a dozen countries are currently developing them. Both the hardware and software are growing exponentially, and this 'double exponential' may bring these developments forward rapidly. In 2015 over a thousand scientists and technology experts wrote a letter warning about this danger, and in 2017 116 founders of robotics and artificial intelligence companies wrote an open letter to the United Nations urging a ban on autonomous weapons similar to that on chemical weapons.

Nanotechnology is a new and exciting frontier in science and technology. It is the study and manipulation of matter in a previously unexplored domain of parameter space, on scales from 1 to 100 nanometres (nm). Richard Feynman planted the first seeds in his 1959 talk *There's Plenty of Room at the Bottom*. The lower limit of 1 nm is set by the size of atoms, and the upper limit approaches the size of cellular life-forms and the scale of microtechnology. Below 100 nm quantum effects become significant and a number of special physical properties emerge, so nanoscales cover an intriguing range (the exact boundary between classical and quantum is debated—in this range classical thermodynamics confronts quantum mechanics in what is now called 'quantum thermodynamics'). Nanotechnology is very broad, ranging from passive nanomaterials that can be used in a variety of applications to the engineering of functional molecular machines using atomic control and inspired by countless examples found in the natural world.

Two developments in the 1980s further stimulated the emergence of nanotechnology: the invention of the scanning tunnelling microscope capable of imaging and manipulating individual atoms in materials, and the discovery of so-called 'buckyballs', naturally occurring molecules comprised of 60 carbon atoms arranged in a hollow sphere, showing what exotic phenomena were possible. Both won Nobel Prizes. A wide range of nanotechnology products started to emerge in the following decades, and today they may number in the thousands. They were initially limited to passive applications such as cosmetics, sunscreen, surface coatings, gecko tape, food packaging, clothing, fuels and nanofibers in tennis rackets and aircraft wings, but may soon include sensors of various kinds, energy storage, medical treatments, nanoelectronics and a host of others.

The possibility of nanomachines ('nanorobots') was discussed by Feynman in his 1959 talk, but remained speculation for many decades (including the 1980s science fiction concept of a 'grey goo' of tiny self-replicating robots consuming everything in their paths). But now they have become a reality. The 2016 Nobel Prize in Chemistry was awarded "for the design and synthesis of molecular machines". The molecular 'nanocar' was selected as one of the ten major discoveries in sciences worldwide in 2011 by the Chinese Academy of Sciences. In 2017 a molecular robot was made at the University of Manchester that can be programmed to build different molecules; it has an arm that can manipulate individual molecules. And a six-speed 'molecular gearbox' was recently made by a team at the University of Tokyo. In the coming years nanotechnology will reach all of us.

The potential benefits are huge, but there are serious concerns about the effects that industrial-scale use of nanomaterials may have on health and the environment. Nanoparticles are pollutants small enough to be absorbed through the skin, and yet they are already in wide use today. Nanofibers, like asbestos, may lead to pulmonary disease if inhaled in sufficient quantities. As is the case with all new technologies, virtually nothing is known about the long-term effects of nanotechnology products. There are calls for tighter regulation, but confusion regarding responsibility. In the meantime there have been proposals for delayed approvals, improved labelling and increased safety information.

Every big step forward brings risks.

4.9 Our Global Impact

A few million years ago our ancestors had no more impact on the environment than any other species; they lived in harmony with nature. In time the early tools enabled humans to gradually inch their way up the food chain, hunting smaller prey while being hunted in turn by larger animals. But the first giant step was the ability to make fire, hundreds of thousands of years ago. Suddenly any human could burn down an entire forest, cooking the resident flora and fauna and changing the landscape to the grasslands that humans prefer. This was the beginning of mankind changing the face of the planet.

Over the seven million years since our ancestors speciated away from the chimpanzees there were over two dozen species of the genus *Homo*, of which ours, *Homo sapiens*, was just one. Now we are the only ones left. What happened to the others? It seems likely that *Homo sapiens* had evolved superior cognitive abilities and language, essential qualities for flexible cooperation in large social groups. While some interbreeding did take place (a few percent of our DNA comes from Neanderthals and Denisovans), it is likely that we either outcompeted them for resources or eliminated them through violence and genocide. Whatever the reason, they became extinct not long after our ancestors turned up: *Homo denisova* upwards of fifty thousand years ago, *Homo neanderthalis* some thirty thousand years ago, and *Homo floresiensis* about twelve thousand years ago. Only our species, *Homo sapiens*, survived.

It wasn't just our cousins that we eliminated. We rose up the food chain and became feared animals in Africa and Asia; the other animals learned to avoid us. But it was a completely different story when we migrated to other parts of the world. There the large native animals had no innate fear of the human newcomers, and could therefore be easily wiped out in a relatively short time. The first such invasion by humans happened in Australia some 50–60 thousand years ago, with the result that almost all of the exotic large animal species were soon eliminated and many smaller species also became extinct. Similar things happened elsewhere when humans arrived. Most of the mammoths and other large animals of North and South America had disappeared by 10,000 years ago, not long after humans made their way via the Bering Strait, got past the Alaskan glaciers and spread out over the Americas. The large animals of Madagascar rapidly disappeared 1500 years ago when the first humans arrived, and the same thing happened in New Zealand when the Maoris arrived about 800 years ago. Waves of extinctions swept over the Pacific islands. Only a few remote islands escaped this fate, including the famous Galapagos Islands off the coast of Ecuador. So it is clear that major human-induced extinctions go back a long time—thousands or even tens of thousands of years. We became the most deadly species on the planet—the "global super-predator".

The Agriculture Revolution that occurred ten thousand years ago independently in various parts of the world caused major change of a very different kind. We began to grow a few species of plants for food. It was good for the plants, but we gave up our free-ranging hunter-gatherer lifestyle for a lifetime of hard labour bending over and tending the plants on small plots, with a much-restricted diet. A loss for the individual but a gain for humanity, whose numbers could swell, leading to villages, towns and civilizations. The bulk of our diet today still comes from plants that were domesticated thousands of years ago—wheat, rice, maize, millet, barley and potatoes. Our ancestors burned down forests and scrubland to make way for agriculture, and these privileged plants were allowed to thrive at the expense of other species, which were considered 'weeds'. Our ancestors also domesticated and bred many species of animals, including sheep, goats, cattle, donkeys, horses, camels, pigs, chickens, dogs and cats. These were used variously for food, clothing, farm labour, transportation and as pets. We, with our entourage of chosen plants and animals, came to be by far the most powerful and dominant species on the planet.

But that was just the start. The serious environmental problems we face today all stem from a few fundamental factors: human overpopulation, rampant development and overconsumption in the 'developed countries'. The human population was just several million 10,000 years ago when agriculture began, 200 million during Roman times, 1 billion in 1800, 2.5 billion in 1950, and 7.6 billion today. The corresponding growth rates were about 0.09% p.a. between Roman times and 1800, 0.6% p.a. from 1800 to 1950, and 1.7% p.a. from 1950 to the present. The rapid exponential growth rate in population over the last couple of hundred years coincided with rapid development in industry and technology, and the combination has had a major impact on the environment.

Food production for our use now involves fully 38% of the land area of the Earth. Two-thirds of that is for pastures, and domesticated animals number in the tens of billions, more than the number of humans and far greater than any wild counterparts. With the advent of the Industrial Age, in just the last couple of centuries we have gone much further in our impact on the Earth and environment. The early factories began to spew toxic chemicals into the atmosphere, railways and roads were built across the landscape, cities rapidly grew, forests were cleared and rivers were dammed. We have now created a massively energy-consuming society based on fossil fuels and electricity for factories, lighting, heating and air conditioning, we have built sprawling cities of skyscrapers and we have manufactured billions of vehicles and aircraft for travel and giant ships for transport. The list goes on and on.

Global energy consumption increased fivefold between 1800 and 1950, and another fivefold in just the last 65 years, corresponding to exponential growth rates of 1.1% and 2.5% p.a. respectively. In 2016 primary energy consumption was comprised of 34% crude oil, 29% coal, 25% natural gas, 7% traditional biofuels, and 6% nuclear, hydro, solar, wind and other renewables. The term "Peak Oil" refers to the maximum rate of extraction, after which it would decline; in the early 1970s it was expected to be reached by 2020, but continued discoveries of new reserves and the advent of fracking have changed the equation. Coal reserves could in principle last for a few more centuries. Energy consumption continues its relentless climb.

The impact of all this is frightening. We threaten the existence not only of other species, but also our own. The classic equation for our environmental impact (I) is $I = PAT$, where P is the human population, A is the per capita affluence, and T is the resource-depleting and polluting technology. What have we done?

Well, for one thing, we are the main cause of what has been called the 'sixth' mass extinction or Holocene extinction, estimated to be producing an extinction rate as high as hundreds or even thousands of times the normal 'background' rate. The previous five mass extinctions took place 443, 360, 250, 200 and 65 million years ago, and each extinguished in the range of 60–95% of all life; two were probably caused by climate change, two were due to supervolcanoes, and the one that killed off the dinosaurs was due to a large asteroid impact. The Holocene is the current geological epoch, which started about 11,700 years ago and happens to cover the history of humanity from the advent of agricultural and the earliest civilizations to the present. It has been estimated that half of all wildlife has been lost over just the last 50 years, and that half of all remaining species could be extinct by the end of this century.

A 2017 paper[6] entitled "World Scientists' Warning to Humanity: A Second Notice" and co-signed by 15,364 scientists from 184 countries states that "we have unleashed a mass extinction event, the sixth in roughly 540 million years, wherein many current life forms could be annihilated or at least committed to extinction by the end of this century". They pointed out that "between 1970 and 2012, the abundance of vertebrates declined by 58%, with freshwater, marine and terrestrial populations declining by 81, 36 and 35% respectively".

That paper was a follow-up to a first warning signed in 1992 by over 1700 scientists including most of the Nobel laureates living at that time. The 1992 authors warned that "a great change in our stewardship of the Earth and the life on it is required if vast human misery is to be avoided", that humans are "on a collision course with the natural world" and that we are "pushing Earth's ecosystems beyond their capacities to support the web of life". They expressed concern about a wide range of issues: ozone depletion, freshwater availability, marine life depletion, ocean dead zones, forest loss, biodiversity destruction, climate change and continued human population growth. The authors of the 2017 paper were able to note some progress over the intervening 25 years: a sharp reduction in ozone-depleting substances, a modest decline in human fertility rates, a slight decrease in the rate of deforestation, and the growth of the renewable-energy sector. But much remains to be done.

---

[6] Ripple et al. (2017).

Biodiversity continues to suffer from a wide variety of factors, including the degradation of rain forests and coral reefs, habitat destruction, overhunting, overfishing, ocean acidification, pollution, invasion by foreign species, introduced diseases, pollinator declines and climate change. The introduction by humans of invasive species of plants and animals into established ecosystems has had huge impacts on the native flora and fauna over large areas. Networks of roads and infrastructure across countries have had major effects on natural habitats and the survival of species, and the degradation and pollution of air, water and soil have been major factors. Pesticides affect far more organisms than just their intended targets, and they can result in widespread land and water pollution. Many disruptions to ecosystems can have subtle and far-reaching effects.

Some of the richest environments on the planet have been severely diminished. Millions of acres of rainforest have been destroyed in order to provide land for livestock; the growing demand for meat has been a major driver of deforestation and habitat destruction. Deforestation not only removes an important carbon sink; it also affects the global climate directly. Coral reefs are the marine equivalents of rainforests, rich in life. But they are dying all around the world because of warming oceans, acidification, overfishing, coral mining, pollution and other factors. Ten percent of the world's coral reefs are already dead, and over 60% are at risk. And large regions of the world have been degraded over the years by pesticides, fertilizers, mining, and various chemical and other industrial pollutants: cleaning agents, aerosols, lead and a host of others. It has been estimated that 40% of the world's agricultural land has been seriously degraded. Non-degradable plastics are notorious waste products of modern society, and it has been suggested that within a few decades there may be more plastic in the oceans than fish. And the availability of fresh water will become a major issue in many countries over the coming years.

But most important of all is climate change, as it has an effect on almost everything in the world. By burning fossil fuels we have increased the amount of atmospheric carbon dioxide by about 40% since the beginning of the Industrial Revolution, and most of that increase has occurred since 1970. Carbon dioxide is a 'greenhouse gas'. As in a conventional greenhouse, the energy from the Sun comes to the Earth's surface as visible light, and is then re-radiated away from the surface as thermal radiation at infrared wavelengths; some of this thermal radiation is absorbed and trapped by greenhouse gases in the atmosphere, causing global warming and climate change. The globally averaged temperature has increased by 0.85 °C over the period from 1880 to 2012.

The Intergovernmental Panel on Climate Change (IPCC), representing the views of more than 2000 scientists, concluded in its 2014 report that "Human influence on the climate system is clear", that "recent anthropogenic emissions of greenhouse gases are the highest in history", and that "Warming of the climate system is unequivocal".

Population and economic growth are responsible for most of the increase in CO_2 emissions. The world population continues to grow, and so does per capita consumption. Where is this heading in the future? The United Nations gives three population projections assuming high, medium and low fertility rates; these predict a population in 2100 of 16.5, 11.2 and 7.3 billion respectively. A simple extrapolation from the current population of 7.6 billion leads inevitably to an increasing population over the coming few decades ('population momentum'). But what population can Earth actually support (its 'carrying capacity')? Guestimates vary widely, as they depend on so many unknown factors, especially the future per capita consumption in the presently developing world (which has the fastest population growth). The UN summarized 65 different estimates in 2012, and found that the most common was about eight billion. So clearly population growth will be a critical factor in any discussions about future development.

Fossil fuel combustion and industrial processes produced about 78% of the total greenhouse gas emissions increase from 1972 to 2010. The primary fossil fuels are petroleum, coal and natural gas. Most of the world's petroleum is used in transportation. Road transportation is the largest cause of climate change, and also has negative effects on air quality, producing smog and acid rain. Rail transportation is somewhat less polluting. Air travel produces particulates and gases that contribute to climate change, and the rapid growth of air travel is a major concern, some calling for a special tax on air travel. Shipping pollutes through greenhouse gas emissions and oil pollution. Petroleum can be toxic to almost all forms of life, and yet it is part of almost all aspects of modern society. Coal is still used in many power plants around the world, and is very polluting. So even aside from climate change, fossil fuels have a large number of negatives.

Climate change is global, and so are its consequences. It has an effect on everything from glaciers, lakes and the ocean to ecosystems, food production and human wellbeing. It has produced a marked increase in the frequency and severity of extreme weather events: heat waves, droughts, floods, cyclones and wildfires, and will ultimately produce long-lasting changes. Ocean warming and acidification resulting from uptake of CO_2 will have increasingly negative effects on marine life and coral reefs. And a large fraction of species face increased risk of extinction because they will not be able to keep up with the rapid pace of climate change.

One of the most obvious effects of global warming is the melting of glaciers in mountain ranges around the world and the ice sheets of Greenland and Antarctica. This, together with the thermal expansion of the warming ocean, results in an increase in the sea level. In the latest IPCC report the sea level rise is projected to be in the range 30 cm to 1 m by 2100, and some estimates put it at 2 m if carbon dioxide emissions continue. The effects are further exacerbated by tides, tsunamis and storm surges from increasingly frequent extreme weather events. This puts upwards of a billion people living in coastal areas and cities, 10% of the world's population, at serious risk. Some of the 130 coastal cities with populations of a million or more may be inundated, and entire countries may disappear (the Maldives, for example, consists of 1100 islands just 1.3 m above sea level). Already coral reefs, beaches and agriculture have been seriously affected in some low-lying islands in the Pacific region and populations have been relocated; Bangladesh is a major disaster waiting to happen.

The IPCC estimates that the global temperature increase from now until the end of the century will be over 5 °C for a scenario with very high greenhouse gas emissions, and less than 1 °C in the case of a stringent mitigation scenario. The now-famous Paris Accord of 2015 aims to hold the global temperature rise this century to well below 2 °C, and strive for an even smaller rise of 1.5 °C above pre-industrial levels. It was signed by nearly 200 countries.

But even if anthropogenic emissions of greenhouse gases were to be stopped completely, many of the effects of climate change will continue for centuries; Much of the damage already done is irreversible on a timescale of hundreds to thousands of years, unless most of the CO_2 can somehow be removed from the atmosphere. This means that sea level rise will continue for many centuries, as will ocean acidification. Other large-scale phenomena such as biosystems, soil carbon and ice sheets have their own intrinsic long timescales.

The "Doomsday Clock" is a concept invented in 1947 by a group of atomic scientists. It is updated annually; it was initially meant to indicate the threat of global nuclear war, but the potential threats from climate change and the misuse of emerging technologies and the life sciences have been included since 2007. It was initially set at seven minutes to midnight; it fluctuated over the years, and reached its most 'optimistic' point at 17 min to midnight in 1991, when the United States and the Soviet Union signed the first Strategic Arms Reduction Treaty and the Soviet Union was formally dissolved. By the beginning of 2018, however, the Clock had reached two minutes to midnight.

Science has certainly contributed to all of the above problems, insofar as it has given us the knowledge (*scientia* in Latin) that is ultimately behind all technologies. Science has therefore been an enabler, and has given us great power. How we use that power—for better or for worse—is completely up to us. Needless to say, we have abused it, especially over the last two centuries, and now we have to face up to the consequences.

But science can also be part of the solution. It is because we have been able to scientifically measure and understand the effects we have had on the environment that we have been able to realize what we have done, to change course, mitigate the effects and in (hopefully most) cases reverse them. There have been several notable successes.

Ozone is one of them. The current concern about the causes of climate change and what to do about it is reminiscent of the debate over the putative link between chlorofluorocarbons (CFCs) and the thinning of the protective ozone layer over 30 years ago. The issue was finally resolved after the discovery of the ozone hole over Antarctica and the strong evidence linking the ozone hole to CFCs. An international agreement was signed to control CFC production, less harmful substitutes were developed, and the ozone hole is now shrinking.

A dramatic example of science, technology and common sense making a huge impact on human survival and well-being was the 'Green Revolution' in the last century. By the late 1960s it was believed that there would be an imminent global food crisis, and in 1968 the American biologist Paul Ehrlich published his book *The Population Bomb* in which he predicted that hundreds of millions of people would starve to death in the 1970s. This, together with the Club of Rome's 1972 best-selling report *The Limits to Growth* and the 1973 oil crisis, produced a great sense of foreboding. But by then the Green Revolution was already well underway. It was the culmination of decades of research and development towards high-yielding and disease-resistant crops, in particular dwarf wheat and rice cultivars, in conjunction with advanced ideas on synthetic fertilizers, pesticides and new methods of irrigation and cultivation. The American agronomist and geneticist Norman Borlaug became known as the 'father of the Green Revolution', was awarded the Nobel Peace Prize in 1970, and was credited with saving over a billion people from starvation.

There are various speculative ideas on how to solve the climate change problem. One involves removing much of the CO_2 from the atmosphere through 'negative emissions': cultivating huge areas of fast-growing plants and trees to extract CO_2 out of the atmosphere and then burning them in power plants, capturing the CO_2 and storing it underground. Another involves directly cooling the atmosphere. It has been noted that powerful volcanic eruptions like that of Pinatubo in 1991 can sometimes temporarily cool the atmosphere. Could this cooling phenomenon perhaps be recreated artificially to permanently counter the warming effect of greenhouse gases? The idea is that a stratospheric aerosol layer is formed by sulphur dioxide gas, which becomes oxidized into droplets of sulfuric acid which in turn reflect sunlight, thereby reducing the amount of energy reaching lower levels and hence cooling them. Such a process (which has an effect on the ozone layer) is considered risky and has long been taboo among many scientists, but there is now support for limited research and readiness for a host of observations of the next big eruption to better understand the possibility of 'engineered cooling'.

As there is now a heightened awareness of environmental issues, both governments and society have shown a willingness to make corrections where possible and modify behaviour to some extent. Many waterways have been cleaned up over the last several decades. There is an increasing emphasis on species preservation. The transport industries are showing greater emphasis on energy efficiency and pollution reduction, and the trend is towards public transportation, increased use of electric vehicles and bicycles. Much waste is recycled, creating new and profitable industries. There is a massive shift underway from incandescent to LED lighting, giving significant energy savings. Much remains to be done, but steps such as these are encouraging. As speakers and authors such as Hans Rosling and Stephen Pinker have pointed out, there are many signs that society at large is moving towards a better world.

And major steps are being taken to wean ourselves off fossil fuels and move to renewable and clean energy sources. The cost of solar panels continues to plummet; Swanson's law (similar to Moore's law in electronics) states that the price of solar modules drops by 20% for every doubling of production; it has fallen from $77 per watt in 1977 to $0.36 in 2014. In several countries solar power is already cheaper than fossil fuel electricity from the grid. Wind power has negligible environmental impact and is compatible with other land uses, but the giant wind farms can be a blight on the landscape, and they will never provide a large fraction of our energy requirements. Hydroelectric power is also clean, but the dams that are often involved can be negative for many species and the environment (although they can also be beneficial for drinking water, irrigation and flood control). Nuclear (fission) power remains only a small contributor to energy resources because of radioactive fuel waste, dangers of nuclear power plant disasters and resulting contamination of very long half-life. The European Union (EU) now gets 30% of its electricity from renewable sources (more than coal), and that fraction will rise to 50% by 2030.

But fusion power is the hoped-for long-term solution for our energy needs. It is the process that powers the Sun. The fuel for reactors will be the hydrogen isotopes deuterium and lithium, both of which can be obtained from sea water, providing energy for millions or even billions of years. And the reaction produces only harmless helium. Fusion is intrinsically safe—it is hard to keep going but if there's a problem it just stops. There is some radioactive waste due to neutrons irradiating the reactor walls, but its half-life is just decades—nothing like the hundreds of thousands of years for waste from today's fission reactors. Fusion has been touted as 'the future' for over half a century, but much has been learned and there are now some very promising prototypes currently being developed, in particular the huge $20 billion ITER being built in France by an international consortium (the EU, China, Russia, the U.S., India, Japan and South Korea); if it is successful the first commercial fusion reactors may come online by 2050. Fusion power would not cause global warming; the world's present energy use is only 0.02% of the solar energy striking the Earth (and only 0.1% even if all countries consumed at the same per capita rate as the U.S.)—totally insignificant compared to sunlight.

The famous "Earthrise" picture on the cover of this book dramatically highlights what a delicate, special and precious planet we live on. We can only hope that it is not too late to correct our course to assure that the Earth and its biosphere will be sustainable as our home into the distant future.

4.10 Science and the Philosophy of Science

It is interesting that there is relatively little communication between scientists and the historians and philosophers of science. Most of the scientists today have never had any courses on the history or philosophy of science; they are totally consumed with the current frontiers of science and their own scientific endeavours.

Originally scientists *were* the philosophers of science. They were called natural philosophers—those who studied the natural world. From Greek times through the Islamic and medieval periods and the Scientific Revolution the promotion of the scientific method was very influential in the establishment of modern science, and the natural philosophers grappled with many of the fundamental issues concerning knowledge and reality. In the nineteenth century natural philosophy became known as science, and the natural philosophers as scientists. What we now know as the philosophy of science then branched off as a separate discipline—philosophers examining the principles and practice of science from the outside.

The first major movement in the philosophy of science in the twentieth century was that of the logical positivists. They greatly admired the rigour and progress of science, and this encouraged them to work towards a new and ambitious 'scientific philosophy' which stressed verification, logic and meaning, and would encompass mathematics, language and science as well as philosophy. Even the possibility of reducing mathematics into a system of logic which could encode all of physics was considered. But these grand and revolutionary ideas were gradually eroded away over the decades by a succession of logicians and philosophers, from logical positivism to a weaker logical empiricism and finally to views expressed generally from within the philosophy of science. The logical positivists, like the general public, saw the progress of science as cumulative, continually creating ever more knowledge. But they regarded the (usually complex) history of events that led to a scientific hypothesis as irrelevant; all that mattered to them was the verification process.

To most scientists the best known twentieth century philosopher of science was Karl Popper, one of the major critics of logical positivism. A strong motivation for him was to reject what he considered to be the unscientific or pseudoscientific claims made, for example by Sigmund Freud. In Popper's view such claims allowed for any possible outcome, and so could neither be confirmed nor denied. His response, as mentioned earlier, was to insist that scientific theories must be *falsifiable*. That is, they must make predictions that can be proven to be false by any observations or experiments that may conflict with them. Falsifiability is an important criterion in modern science.

Popper also agreed with the eighteenth century philosopher David Hume that theories based on induction (which includes most of science and everyday life) can never be absolutely proven. Induction involves an extrapolation. A theory based on a limited sample of the natural world can never be proven to apply to all samples, no matter how many experiments or observations support it. Popper tried to conceive of a scientific method that involves only falsification, which would be truly deductive, resulting in theories that can be proven to be absolutely true.

Thomas Kuhn is considered to have been the most influential historian and philosopher of science over the past century. His book *The Structure of Scientific Revolutions* (1962) changed the way philosophers viewed science. Until that time, the philosophy of science had 'sanitized' science by disregarding everything that preceded a hypothesis as irrelevant, and examining only the justification process. The cumulative nature of science was taken for granted. Kuhn took an orthogonal view, in which history plays the dominant role, and science advances not smoothly, but in revolutions.

In Kuhn's simple model, science proceeds in a series of revolutions, between which there are periods of relative calm in which 'normal science' can be done. Each of these periods is characterized by a prevailing 'paradigm'; it is the current worldview of the scientists, and determines what they work on and how they do it. He called their work during these periods 'puzzle-solving', by which he meant that, as in the case of newspaper puzzles, it is virtually guaranteed (by the existence of the paradigm) that they have a solution. The science is done in the context of that paradigm, and the scientists must not question the paradigm or think beyond it. In this way, according to Kuhn, 'normal science' is efficient and cumulative.

Over the course of time, 'normal science' starts to reveal anomalies in the paradigm. At first these are assumed to be due to mistakes by the scientist who encounters them. But as their numbers increase, it becomes apparent that the paradigm itself has problems. A crisis develops. At this stage alternatives to the paradigm are proposed, and eventually one is found that looks the most promising. A battle ensues between the supporters of the old and new paradigms. It is a revolution. The paradigms may be 'incommensurate', with different values and standards, and the opponents 'talk through' each other. Ultimately the new paradigm has the most support, and the revolution is over. The majority of scientists move on to 'normal science' in the context of the new paradigm, and those left behind either leave the field or die off. This cycle is repeated over and over again.

Kuhn's model seems stilted and artificial. In some cases he forced the facts to conform to his model. He took extreme positions in some areas. He obviously loathed science textbooks for 'disguising' the true history of science. His 'crises' extended over decades, centuries or even millennia, rather than years. He felt that there may be no net growth of knowledge after all the revolutions. His command of history was impressive, but his examples were confined largely to Copernicus, Newton, Lavoisier and Einstein—hardly a mention of Darwin, Faraday or Maxwell. Nevertheless his book had a huge impact, selling upwards of a million copies and being included in lists of the best books of the last century. It was very influential not only in the philosophy of science but also in sociology and other areas of the humanities. Immediately after publication there were harsh criticisms and heated debates, and Kuhn spent years softening some of his more extreme positions. His most enduring contribution was the popularization of the terms 'paradigm' and 'paradigm shift'.

To practicing scientists Kuhn's model can seem rather quaint and foreign. Certainly there are revolutions and paradigms of all magnitudes, but science is vastly more, as elaborated above in Chap. 3. Scientists are not at all confined to their paradigms—they strive to see 'outside the box', and would love to be the first to make the next big change in science. And discovery plays a major role in science, while it hardly features at all in Kuhn's model. In spite of its having had a huge impact on philosophy, Kuhn's book has not influenced the course of science.

In the 1960s and 1970s a rebel by the name of Paul Feyerabend burst onto the scene. He reacted against Kuhn's notion of 'normal science', which he saw as restricting the freedom of scientists, and he criticized the scientific method itself for the same reason. He felt that Kuhn's paradigms could not restrict scientists, as they are always on the lookout for new ideas. His 1975 book *Against Method* introduced the term 'epistemological anarchy', rejecting any rules and constraints on science and leaving scientists free-thinking, creative and opportunistic. His approach has been characterized as 'anything goes', implying that, in his view, there can be no strict definition of science.

One of the significant debates in the philosophy of science over the past century has been that between scientific realists and anti-realists. The realists maintain, for example, that the subatomic entities such as electrons and muons are physically real, whereas the anti-realists think they may be merely useful fictions, enabling hypotheses, predictions and further discoveries to be made. The realist position accords with the common-sense view that science is discovering real things that are 'out there'. The anti-realists hold that they may or may not be real—there is no way of deciding, so their position is one of agnosticism.

Arguments have been made for both of these positions. The realists contend that, as the theory containing electrons and muons has been extremely successful, it would be a remarkable coincidence unless those entities do indeed exist. But the anti-realists have pointed out that there have been apparently successful theories in the past positing things such as phlogiston, caloric and the aether, which have turned out to be false. Another argument from the realists is that it is not possible to draw a hard line between the domain of the unobservable and that of the observable. There is a continuum, so it is as reasonable to regard electrons and muons as real as hard-to-see bacteria and trees. The anti-realists have argued that the theories positing these entities may be non-unique; there may be many other possible theories that could equally well explain the data. The realists then point out that successful theories are rare, and that scientists can have difficulty finding even one that fits the data. The debate continues.

Some working scientists have also taken the time to ponder the fundamentals of science—questions that would have interested the ancient Greeks. An outstanding example was Einstein who, even though he was one of the founding fathers of quantum mechanics, later became very disturbed by the implications (unlike many other physicists whose motto was "shut up and calculate" because the predictions of quantum mechanics were so correct and precise, and there was so much to be done). Quantum uncertainty implies that the subatomic world is ruled by probability, and Einstein once said "God does not play dice". He felt that there must be some deeper underlying reality in which strict cause and effect determine events. He had many frustrating discussions with Neils Bohr in the 1930s. In exasperation, in 1935 he and two colleagues invented a 'thought experiment' which seemed to contradict the predictions of quantum theory in principle, although they never thought that any such experiment could ever be carried out. But in 1963 the physicist John Bell came up with an ingenious version of the experiment that was conceivable, and in the next decade it was carried out. The quantum theory was proven correct and Einstein was wrong. That result, inspired by Einstein's persistence, actually turned out to be an important step forward in physics. Nevertheless, and in spite of recent stunning successes in quantum experiments, worries still persist about the fundamental implications of quantum mechanics.

It is certainly important (as well as interesting) for both scientists and philosophers to contemplate the fundamentals of science. And the views of some philosophers such as Karl Popper have certainly been influential in guiding the way many scientists approach their work today. Closer interaction between scientists and the historians and philosophers of today would undoubtedly be beneficial.

4.11 Science and Religion

Science and religion were destined to clash. In explaining the world they are antithetical—the religions of the past attempted to explain events in terms of a pantheon of supernatural beings, whereas science succeeds in explaining events in terms of laws and causes that are part of the natural world itself. The previous two hundred pages of this book were devoted to the origin, development and nature of science, and now, for this section, we also have to know something about the origin, development and nature of religion.

One early motivation for religious belief was undoubtedly the mystery of death. Some other species of animals also appear to mourn their dead, but our ancestors seem to have taken this a step further. There is evidence from as long as a hundred thousand years ago of human funeral rites. As the living breathe but the dead do not, one might think that something vital left the body upon death: a supernatural 'soul'. In that case perhaps this 'essence' of the deceased person still exists somewhere—in an afterlife. Perhaps we all end up in that place, together again. Ornaments and tools have been found in some very ancient graves, perhaps intended to be used by the deceased in the afterlife. This practice evolved over the millennia, and resulted in some spectacular examples that we can still see today, such as the Great Pyramids of the Pharaohs in Egypt and the buried Terracotta Army in Xi'an to protect the tomb of China's first emperor.

The other great mystery for our early ancestors would have been the very fact of the existence of life, the world and the heavens above. Where did all this come from? Perhaps some all-powerful being created everything, including space and time. And perhaps this supernatural being also watched over and protected humans, and even communicated with them from time to time. As the world is a complicated place, with wind, rain, floods, droughts, lightning and thunder, perhaps there were also lesser gods responsible for these phenomena. The pantheons of religions would eventually include spirits, devils, demons, angels and myths. There could be hundreds of gods and thousands of myths in some of these 'polytheistic' religions, which appeared in virtually all cultures around the world. There have probably been thousands of these religions over the course of human history, each of which was fervently believed in its time, but almost all of them have disappeared.

Of the major religions today, Hinduism is certainly the most reminiscent of the early polytheistic religions. It dates back thousands of years, and stems from a diversity of influences resulting in a rich and complex culture. According to Hindu scriptures there are hundreds of millions of gods, although they might all be considered expressions of one God. A central theme of Hinduism is the wheel of life: we all live in a continuous cycle of birth, life and death. In death we are reborn to live another life (and not necessarily as a human—one may come back as any species). How one lives in this life and previous lives (one's *karma*) determines how one will live in the next. Final release from the wheel of life remains the ultimate goal.

Siddhartha Gautama (the Buddha), born in 580 BC, was appalled by the doctrine of endless cycles. Various experiences and insights led him to the conclusion that desire was the cause of human suffering. Then, famously contemplating under a wild fig tree, he was able to rid himself even of desire itself, and went into a state of ecstasy—he was released from the wheel of life and became The Enlightened One. On hearing of this a group of monks became his followers, and, although Buddhism is more a practice than a creed, it became a major religion in East Asia.

Confucius was born in 551 BC, at a time of upheaval in China. In contrast to the anger and violence that were raging throughout the country, he proposed that the warlords instead focus on the welfare of the people. He felt that the world should be managed ethically, and he encouraged compassion and the development of skills in handling disagreements. We should value society above the desires of the individual. Courtesy and respect were emphasized. It was a philosophy for the betterment of daily life. Of course it acknowledged the reality of death, but it considered that the dead had not ceased to exist—rather, they have a continuing presence in our lives and communities, and are to be revered. Confucianism is largely a philosophy of life, but it is also a religion.

Another major religion in China is Taoism. Lao Tse was born around 600 BC, and he developed a different approach to life. Confucius put the good of society above that of the individual, but Lao Tse stressed the freedom of the individual. He admired the wholeness of nature, and felt that humans should be a part of it. There is peace in harmony. He disliked the rules and organization of society, and encouraged being part of the smooth flow of nature. He believed that balance in all things is important, and introduced the complementary concepts of Yin and Yang. Taoism is also more than a philosophy, and has many gods. Humans can become immortal gods through meditation and the suppression of human desires.

The religions that originated in the Middle East were very different. Abraham, who lived in Mesopotamia around 1800 BC, introduced the shift from polytheism to monotheism, and is considered to be the founding father of Judaism, Christianity and Islam.

Abraham was a prophet. There have been hundreds of prophets throughout the history of religion. These are the 'chosen ones' who claim to have had a vision or mystical experience and then announce it to the world, starting a new movement or religion. According to stories passed down to us, Abraham was upset by the polytheism of Mesopotamia, especially the worshipping of idols. Eventually he heard the voice of God telling him emphatically that there is only one God, and that he should leave this land of idol-worshippers. He dutifully set out with his family and flocks and settled in the land of Canaan (modern-day Israel and Palestine). Over the ensuing millennia they faced several challenges but prevailed. However, in 63 BC the Romans took over, and in the year 70 AD they dispersed the Israelites (Jews) to the four corners of the Earth for a two-thousand-year exile.

The story of the prophet Jesus of Nazareth has had a huge impact on religion and world history. Virtually all scholars agree that such a person did in fact exist, but beyond that the historical accuracy of the various accounts written by his followers is rather uncertain. The stories of his birth, his compassion and his miracles are famous. In addition to his twelve disciples, he attracted a large following. Was he the Messiah (Christ) that Judaism had been anticipating for years? Was he the Son of God? The Jewish religious leaders saw him as a pretender, and the Romans saw him as trouble-maker in a difficult part of the empire. His crucifixion and the subsequent tales of his resurrection became legendary, and Christianity was born. It spread widely, and the Romans persecuted Christians throughout the empire, but in a stunning reversal Christianity became the official religion of the Roman Empire in the year 380.

Another prophet emerged centuries later, in the Arabian peninsula. Muhammad, born in Mecca in 570, started out as a camel driver and became a successful trader. But he became disillusioned by the corruption and idolatry he saw around him. He went to pray in a cave, and he had visions and heard voices coming to him via an angel. In 613 he began to preach in Mecca. He and his Muslim followers had to fight against the establishment, but they prevailed. Muhammad died in 632, but his struggle continued. The message was simply "There is no god but God, and Muhammed is the prophet of Allah". The last part was meant to declare that there could be absolutely no more prophets. But of course there were.

Over the course of history there were schisms in the major religions. Not long after Muhammad's death disagreements over succession caused Islam to fracture into two groups, Sunnis and Shias, which have been violently opposed to each other ever since. In 1054 lingering differences between two versions of Christianity caused the Great Schism: the Catholic Church in the west and the Orthodox Church in the east. In 1517 Martin Luther, realizing that believers could receive the word of God directly from the newly printed Bible rather than through the corrupt clergy, started the Protestant Reformation in Wittenberg, Saxony; Protestants and Catholics have lived different lives ever since, and there have been many violent clashes between them, and between Protestants and other Protestants. Another schism with the Catholic Church was caused by England's Henry VIII, who wanted a new wife; the result was the new Church of England.

Protestantism became a rich new source of prophets. In 1648 George Fox in England had a revelation and heard the voice of God. He said there was no need for organized religion at all, and claimed that all humans are equal—male or female, slave or free. Thus was born the Quaker movement; they escaped persecution by going to America. In 1830 Joseph Smith in upstate New York had a vision; an angel told him to avoid the corrupt local churches, and to look for buried golden plates bearing the writings of the prophets of ancient America. He reportedly found them, wrote the Book of Mormon and established the Church of Latter-Day Saints. He was murdered in a prison, but then Brigham Young led the Mormons on a long trek to Utah, where they established themselves.

The ultimate in prophetic religions is the Bahá'i faith. It was established in Iran in the mid-nineteenth century by a man who called himself the Báb and said he was a herald from God preparing the way for a forthcoming prophet. As Islam had declared that there could be no further prophet following Muhammed, the Báb was arrested and summarily executed. A few years later a man by the name of Bahá'u'lláh said that he was the expected prophet, and the Bahá'i faith was born. Bahá'u'lláh was imprisoned, exiled, and died in a Palestinian prison in 1892, but the faith continues. It is the idea of progressive revelation, as successive prophets, hearing from God, take their turn.

Over the millennia countless religions, gods and myths have come and gone. We are left with just several major religions today, and a few hundred minor ones. What are we to make of all this?

One thing we can be sure of is that religion has been a major cause of violence and war throughout history. This is certainly true for the three Abrahamic religions. It has been said that more blood has been spilt in the name of religion than for any other cause in history. Because religions are based on faith alone, their claims and creeds are taken as absolute—the one true faith. Therefore, when religions clash it can be a fight to the death. History is full of such violence, between Christians and Jews, Christians and pagans, Christians and Muslims, Sunnis and Shias, Catholics and Protestants, Muslims and Hindus, Buddhists and Hindus. Today the mainstream religions are largely peaceful, but there are still pockets of violent religious extremism.

Science and religion are totally different in the ways they 'explain' the world. Some religions have rigid, dogmatic beliefs that cannot be questioned, whereas science has theories that are forever subject to experimental or observational tests. So if such a religious faith is questioned, the reaction can be violent. But if a scientific theory is questioned, an objective test can be made to check its validity—everyone knows that the ultimate arbiter of a scientific debate is nature itself.

Most scientists in the past were religious, and some still are today. A common theme over the ages was that science was unveiling the marvels and laws of nature that God had put in place. But there has been a long history of interactions between religion and science, most of them antagonistic. Many of them were mentioned in Chap. 2, but in the following few paragraphs they are summarized together in order to highlight the history of the relationship between science and religion.

Religion predated natural philosophy by tens of thousands of years. In the early civilizations, as we saw earlier, the worldviews were totally dominated by religions, gods, spirits and myths. The atmosphere in those early monolithic and religious civilizations was not conducive to free and rational thinking and independent contemplation of nature.

Ancient Greece was a very different kind of place, being comprised of independent city-states. There was religion, but it was fragmented and there was no overall priestly caste to impose dogma. Citizens were free and debate was customary, so it is not too surprising that natural philosophy took root there. It started with the revolutionary idea that the causes of events are part of the physical world itself, and can be studied by rational thought with no recourse to religion or mythology. Greek philosophy reached its zenith in about 300–400 BC, and then slowly declined. The causes of its demise have been much discussed, but the rise of Christianity was undoubtedly a major factor in the late stages.[7]

When the Islamic empire emerged in the seventh and eight centuries there was only one text in Arabic, the Qur'an, which encouraged learning and knowledge. The result was a massive 'translation movement' of the Greek classics into Arabic, and this led to the rise of Islamic science in the tradition of the Greek natural philosophers. It lasted for some 400 years—the 'Golden Age of Islamic Science'. Its decline was due to a reaction against non-Islamic influences; books were burned and the madrasses restricted their curricula to the Qur'an.

Christianity had a positive influence on natural philosophy for a while during the Dark Ages and early Medieval Period: some of the Greek classics were collected, preserved and translated into Latin in many monasteries. And many more were translated by various scholars from Arabic and Greek into Latin. So when the first European universities were established in the eleventh and twelfth centuries, their curricula were comprised largely of the Greek classics.

Aristotle was the most influential Greek natural philosopher over the years, and by the thirteenth century the Roman Catholic Church became concerned about the obvious conflict between his worldview and the dogma of the Church. Aristotle considered the universe to be eternal; the Church held that it had been created by God. Aristotle thought that events were determined by natural cause and effect; the Church held that God could cause events by divine intervention and miracles. Aristotle's geocentric crystalline spheres would get in the way of the biblical Ascension into the heavens. The Church banned the reading and teaching of Aristotle's works at the University of Paris for years on pain of excommunication. The scholar Thomas Aquinas struggled

[7] Nixey (2017) The Darkening Age: The Christian Destruction of the Classical World.

with these issues and was initially strongly opposed by theologians, but eventually he found a compromise, and 'Thomism' became the official position of the Roman Catholic Church.

Even the Protestant movement of the 1500s was hostile to rational thought: Martin Luther emphatically proclaimed several times that reason is the greatest enemy of faith.

In the sixteenth century Copernicus proposed his heliocentric model of the cosmos. He knew that it was contrary to the geocentric and anthropocentric worldview of Aristotle, which had previously been adopted by the Roman Catholic Church as part of 'Thomism', so his work was not published until he was on his deathbed. Decades later the heretic Giordano Bruno was burned to death at the stake by the Church, in part for his support of the heliocentric model. In the early seventeenth century Galileo Galilei made his famous astronomical discoveries which strongly supported the Copernican view. He published two books on this, one of which was particularly incendiary. His books and that of Copernicus were banned by the Church. Galileo was forced by the Roman Inquisition to renounce his beliefs, and condemned to lifelong house arrest. The Roman Catholic Church finally pardoned him in 1992, 350 years after his death.

At about the time of Galileo's trial, Archbishop James Ussher in Ireland calculated from the Bible that the year of the creation was 4004 BC. In the nineteenth century the age and history of the Earth was a major issue, and there were two contrasting views. One was catastrophism, according to which the Earth was moulded by catastrophic events such as the biblical flood. The other was uniformitarianism, according to which the Earth has been shaped by long-term geological processes such as those we see today. Both uniformitarianism and Darwin's theory of evolution implied that the age of the Earth would be far greater than previously thought, in contradiction to the 6000 years since the biblical creation. It is now known that the age of the Earth is 4.6 billion years.

Over the course of time the harsh realities of the world of life were becoming clear to naturalists. They came to realize that life has to feed on life—that carnivores and herbivores have no option but to eat other living things in order to obtain the energy and organic molecules required for life. It is of necessity a very cruel world—"red in tooth and claw", as Tennyson famously put it. How could a loving God have created such a monstrous world? This was a very disturbing reality that those believers had to face up to.

Following his famous voyage on the *Beagle* in the early nineteenth century, Charles Darwin wrote an outline of his ideas on evolution by natural selection. He kept it to himself because he did not want to upset his wife who was quite religious, and he was not yet prepared to expose his ideas to his contemporaries and the Church, which he knew would have a hostile reaction. It was one thing to suggest that evolution could take place in the world of life at large, but the thought that we humans could be descended from the apes would be too much for many (in spite of the obvious anatomical similarities—chimp DNA is 98.5% the same as ours). Darwin finally published his masterpiece *On the Origin of Species by Means of Natural Selection* in 1859. Because it was so thorough and compelling, it was immediately acclaimed, but it also drew the expected attacks.

In 1860 the 'Great Debate' took place in Oxford. A friend of Darwin's, Thomas Huxley, supported the evolutionist view, while Samuel Wilberforce, Bishop of Oxford, upheld the concepts of biblical creation. Not surprisingly both sides claimed victory, and the debate has continued up to the present, with periodic outbursts such as the 1925 Scopes Monkey Trial held in Tennessee. Over the last decades the religious view has come in the guise of 'intelligent design', which purports to present an alternative to Darwin's theory of evolution. As it turns out, the mechanism for Darwin's theory was already being worked out by Gregor Mendel at about the same time that Darwin published *Origins*, and the ultimate confirmation of Darwin's theory has recently been provided by something he could never have dreamt of—molecular genetics: the sequences of mutations in the genomes perfectly match the evolutionary history from Darwin's theory of evolution by natural selection.

It is very striking that science arose only once in world history, and managed to survive its precarious passage as a thin thread from the ancient Greeks via the Islamic and medieval epochs to the Scientific Revolution and modern science. The fact that there were already pre-existing all-encompassing religious worldviews firmly in place very early in history probably helps to explain why natural philosophy did not arise in other parts of the world. Science was pre-empted by religion, and the religions became very intolerant of free thinking. World history could have been very different.

But science did survive, and there is no question that it has over the years replaced religion in explaining a wide range of experiences in the natural and physical world. The terrifying and mysterious phenomena such as lightning, thunder, rain, floods, droughts, earthquakes, eclipses, comets, meteors, the Sun, the Moon, the planets and the stars that prompted ancient cultures to invoke a variety of gods, myths and prayers are now explained by science. Not only can we now explain those phenomena in terms of laws intrinsic to nature, we can also use those laws to make a huge number of successful predictions of future events in the world, in many cases to astonishing degrees of accuracy.

Indeed, science has been so successful that the number of significant questions left in our easily observable world, in which religion could still claim to provide the only answer, is very small. The Church has had to accommodate the continuing progress of science, and no longer takes issue with scientific realities such as the universe, the nature of matter and the basis of life.

So it is interesting to imagine a world in which religion had never existed, but in which people live as we do now in a world in which science has explained all that it has. What would now motivate people to begin to have religious thoughts, beliefs and prayers? After all that science has explained, what issues would be left to motivate religious belief?

One would undoubtedly be the fear of death. That is understandable, but death itself is no mystery. We know what happens to the body. The issue is obviously about the possible existence of a 'soul', and 'life after death'. Virtually all neuroscientists agree that the soul is nothing more than the consciousness we experience, and that consciousness itself is purely a function of the material brain. If that is the case it cannot exist without the brain, and simply ceases to exist when the physical body and brain die. We are learning more about consciousness every day.

Another is the question "What is the purpose of life?" This is not a meaningful question for science, which works from past causes, not towards future goals. If it means "What is *my* purpose in life?", then that is just a personal matter. If it is meant to imply the existence of a purpose-giving agent, then again science has nothing to say on the matter, aside from the fact that there is no scientific evidence for such an agent.

The origins of life and the universe have been at the core of most if not all religions over thousands of years. Both are valid topics for science, and considerable progress has been made. We know that life on Earth began between 4.6 and 3.5 billion years ago, and the origin of life is an active field of study; it will be solved if we can eventually understand the processes that can lead from abiotic chemistry to life, or if some sort of artificial life is actually produced in the laboratory. We now know that there are billions of Earth-like planets in our galaxy, and probably billions of trillions in the universe; our planet Earth is probably not unique, and it is hard to imagine that we are alone in the universe. The beginning of our universe is known to have taken place 13.8 billion years ago, and its evolution is now well known. Some scientists believe that even its origin is now understood by science; others would not go that far, but in any case science has certainly been successful in unravelling many of the mysteries of the very early universe.

Another family of subjects that are sometimes mentioned are paranormal and supernatural phenomena. If they are literally 'supernatural' then of course science has nothing to say about them, as they have nothing to do with the natural world we live in. But any phenomena that have any effect whatsoever on the material world (including the brain), are certainly potential subjects for scientific investigation. So far no such phenomena have become part of the body of rigorously established scientific knowledge.

There is one rather striking line of thought originating from modern science that may be considered relevant to religious belief, and which is not widely known. As discussed in the next chapter, there are several 'coincidences' in the large-scale properties and physics of the universe which appear to be essential for life as we know it. It seems as though our universe is 'fine-tuned for life'. For some scientists this is explained by the 'anthropic principle': we can only exist in a universe that has the right conditions for life. If there is a vast multiverse of different universes (as many scientists think), it is no surprise that our universe is one of the ones with the right conditions, as otherwise we would not be here. But any individuals of a religious bent may be inclined to interpret this in terms of a God who designed the universe especially for us. The debates about the anthropic principle and the multiverse continue to the present day.

There is no question that the vast majority of believers see their religion as peaceful and respectful of other religions, in spite of the current hotspots and terrorist attacks. And many of the religions of the world today serve a number of positive cultural functions. There are large regions of the globe in which religions help to maintain stability, promote good values and do good works; they can provide a sense of comfort and security for individuals, venues and ceremonies for major life events (birth, marriage and death), and support for social communities.

Secular societies and humanism may eventually replace religion in many of these functions. Neither morality nor the organization and running of a civilized society requires religion, and humanist celebrants can lead the events related to birth, marriage and funerals.

It is interesting to compare religious attitudes in different parts of the world today. Americans are far more religious than Europeans, and religious belief in most European countries continues to decline. Polls indicate that 98% of Indonesians, 80 % of Indians, 60% of Americans, 21% of Europeans and 3% of Chinese consider religion to be 'very important'.

Like most people, scientists rarely discuss their religious beliefs, if any. Newton ended his *Principia* by praising God for His creations. When the great scientist Pierre-Simon Laplace was asked by Napoleon why he did not mention God in his new book, he replied "Sire, I had no need of that hypothesis". Darwin was concerned that his theory might offend the religious sensitivities of his wife. For Einstein, God is simply everything that exists, and plays no role in everyday life (a view called pantheism).

But in recent years many of the leading scientists of the world have been polled about their religious views. A 1998 poll[8] of members of the National Academy of Sciences in the U.S. found that only 7% and 8% of the scientists believe in God and immortality respectively. In contrast, about 95% of the American population at large believe in God and over 70% in immortality. A similar poll[9] was also taken of the Fellows of the Royal Society of London in 2013, and again only 8% of the scientists expressed a belief in God and in consciousness surviving death.

Biologists tend to be more atheistic than physicists, and they tend to be less likely to accept that science and religion are compatible. Both tendencies have been ascribed to the greater exposure to the raw issues of evolution and life, and the fact that the biological sciences today bear the brunt of religious and social interference in such matters as genetics, cloning and stem-cell research as well as evolution (still the favourite target of creationists and proponents of intelligent design).

Some scientists prefer to be effectively outside of the theist-agnostic-atheist continuum[10]—they don't have a 'position' and simply don't care at all about religion. They are just preoccupied with their life in the real world of science, and religion is not an issue for them.

Science continues to explain more and more phenomena in the natural world as time goes on. Religions have typically been reactionary, and even today we can see a continuing clash between the Church and liberal views on many subjects. But science remains neutral and unbiased; it is just concerned with objective knowledge about the natural and physical world.

[8] Larson and Witham (1998) Leading scientists still reject God.

[9] Stirrat and Cornwall (2013) Eminent scientists reject the supernatural: a survey of the Fellows of the Royal Society.

[10] According to some definitions, agnosticism is as much a belief ("unknowable") as are theism and atheism.

4.12 The Massively Interdependent World

It has been said that Thomas Young (1773–1829) was "the last man who knew everything". He was a polymath and physician who made important contributions to the fields of optics, mechanics, energy, physiology and Egyptology. In addition to his native English, by the age of fourteen he knew Greek and Latin and was acquainted with eleven other languages. He is best known for having established the wave theory of light. In the Encyclopaedia Britannica he compared the grammar of 400 languages. He contributed to the deciphering of the Rosetta Stone. He was certainly impressive, but even he would be completely overwhelmed by the world's knowledge today.

Today our vast scientific knowledge is shared amongst all of us. Any individual can only know a tiny fraction of the world's knowledge; there are experts in every conceivable niche, and we rely on them for their knowledge.

But now, thanks to the Internet, we can all access the world's entire knowledge wherever we are—like the Great Library of Alexandria, but far greater and more accessible. Whatever the field, however small the niche, it is accessible (with some exceptions, such as secret defence or private industrial research and copyright material). The Internet is probably the greatest advance the world has known over the last half century.

One might consider the Internet to be merely a *quantitative* development—just more of the same that we've known for decades. But over the last 10–15 years the exponential growth and stupendous power of the Internet seems to have surpassed some magical threshold. An 'emergent property' is a property of a complex system that the entities comprising it do not have—like the 'wetness of water', or the 'temperature of a gas'. Consciousness is sometimes referred to as an emergent property of the brain. A big enough *quantitative* increase can manifest itself as a *qualitative* change. So perhaps we could regard the Internet as an emergent property—a fundamentally new phenomenon.

One of the miracles of the Internet is the free online publicly editable encyclopaedia *Wikipedia*. It was established in 2001, and by 2018 it contained over five million articles in English (40 million in all its 293 languages) and has 33 million users. A peer review of science articles appearing in both the *Encyclopaedia Britannica* and *Wikipedia* was published in *Nature* in 2005,[11] and found their quality to be quite similar. Volunteers around the world have self-organized to care for the articles in their areas of expertise.

[11] Giles, J. (2005) Internet Encyclopaedias go Head to Head.

But, even aside from *Wikipedia*, search engines can find articles, documents and entire courses online. There are countless websites accessible, so one can go straight to the source. This has been a revolution for billions of users. It is worth remarking that the World Wide Web was invented by Tim Berners-Lee in 1989 at CERN—another offshoot of pure science.

The Internet is now central to almost everything, but it is just one of the miracles of modern life. Through science and technology we have built a multi-dimensional world that is so massively interdependent as to be almost incomprehensible.

Consider the humble pencil. Neither you nor I would be able to make it by ourselves. It is made of four parts (the graphite core, the wooden shell, the eraser, and the metal tube that holds the eraser in place). These all come from different places. The technology and machines that make each of them are different, and those machines in turn were designed and made by hundreds of others. The workers who make one part do not know the workers who make the other parts.[12]

Look around you at the many simple things you see—the hammer, the teacup, the mattress, the zipper, the kettle, the scissors, the chairs, the door-knob, the lightbulb, the desk, the ladder—all of these were made by different people in different places using different machines and different raw materials.

At the other extreme is the 747 jumbo jet. It has some six million parts. These come from all over the world, and require many independent manufacturing technologies and different raw materials. A car has about 30,000 parts, down to the smallest screw, and again there are many independent companies that make and supply these parts.

The worker who helped to make a 747 wears a shirt that was made by someone who is a passenger on that plane. They don't know each other—they live on opposite sides of the planet. The entire modern world of billions of people is massively interconnected, with workers being both producers and consumers. The typical car owner has no idea what most of the parts of the car are, and certainly no idea of where they came from. It has been said that today, 80% of global trade comprises international supply chains, and parts of many products criss-cross several borders. Globalization is increasingly a tightly-woven fabric.

Our complex interdependent world of science and technology is something like the amazing enterprises of the superorganisms—the ants, the wasps, the termites and the bees. Every individual contributes, with no knowledge of

[12] This example comes from the excellent essay entitled "I, Pencil", written by Leonard Read (1958).

what most of the others are doing. And miraculously it works, to the advantage of all.

Actually, our interdependent world is far more complex than that of the superorganisms, as there are so many more degrees of freedom. In their case there are only a limited number of types of individuals—queens, and workers doing a finite number of specific defined tasks—whereas humans around the world carry out an almost unlimited range of diverse activities. In our case it works because, while each individual is following his or her self-interest (guided by Adam Smith's 'invisible hand', earning money in the process), they are ultimately contributing to the common good. Something so complex cannot be organized 'top-down'—it requires the freedom of the individual.

We would be totally lost if this huge edifice of complex technological and commercial enterprise suddenly came crashing down. Our modern world requires constant attention, repair and replacement in order to continue to function. A power outage of several days could wreak havoc. Supermarket shelves have to be re-stocked on a daily basis. The entire global system is increasingly locked together as one, so it is vulnerable to shocks such as the global financial crisis of 2008. There will soon be over 20 billion devices connected to the Internet. With such dependence on technology we have gone way out on a limb, and our interdependent world has become highly vulnerable. We must manage it with great care.

5

Into the Future

5.1 Will the Current Pace of Science Continue?

The brief overview of the history of science in Chap. 2 revealed several distinct periods: The extremely long period of slow Palaeolithic development, the rise of towns, cities and civilizations, the 'Greek Miracle' that introduced the concept of causes in nature, the Islamic and medieval periods, the Scientific Revolution that arose from the Greek tradition and established modern science, and the recent few centuries of exponential growth in science. We are certainly at a high point in the history of science right now. But will this exponential pace of growth continue into the future?

The notion of progress is central to our worldview today. We expect new science and new technologies all the time: improvements in everything everywhere on a regular basis. This is a very new phenomenon. Will it always be so in the future? It certainly wasn't in the past. Most of our ancestors eked out a living day to day, hoping merely that tomorrow would be like today.

Science is now an integral part of the very fabric of our society. Science and technology are closely intertwined, and together they have given us a living standard far higher than could ever be imagined in previous centuries.

While the roots of modern science are European, it is rapidly being adopted by all the cultures of the world, as its benefits are so obvious. The most famous adaptation to 'Western Science' was the Japanese Meiji Restoration of the late 1800s. Within decades Japan went from an agricultural society to a modern industrial one; by the 1940s it was able to compete militarily with the Western powers themselves, and by the 1960s it was one of the major economic powers

P. Shaver, *The Rise of Science*, https://doi.org/10.1007/978-3-319-91812-9_5

of the world. China's rise is even more remarkable; it emerged from the devastation caused by the Cultural Revolution in the late 1970s, and its GDP is now 60% that of the U.S. Other rapidly developing countries are making the transition today. As a result, science itself is becoming a worldwide endeavour, with more and more scientists and engineers adding to the pool of talent, and the growth of science is further enhanced.

It is true, however, that we have already come quite a long way since the Scientific Revolution, and some of the 'big questions' of the past have now been answered, as shown in Chap. 2. We have explored far beyond the solar system, right back to the beginning of the universe 13.8 billion years ago. Newtonian and Einsteinian physics explain events on the largest scales. We have found the atom and studied physics on scales millions of times smaller. Quantum mechanics and the Standard Model of particle physics explain events in the atomic and subatomic world, underpinning electricity, chemistry and modern electronics. We understand continental drift. We have discovered the genetic basis of life and explained how life evolves. Several of the big questions seem to have been 'wrapped up' by the end of the twentieth century. With monumental progress like this, one may wonder how much is left to do.

But similar views were expressed in the late1800s. Students were advised against going into physics, as everything had already been done. In 1874 Philipp von Jolly, a professor of physics at the University of Munich, advised Max Planck against going into physics, saying that "Almost everything is already discovered, and all that remains is to fill a few holes". In 1894, Albert Michelson said that "The more important fundamental laws and facts of physical science have all been discovered…Our future discoveries must be looked for in the sixth place of decimals", and in 1900 William Thomson (Lord Kelvin) is said to have proclaimed that "There is nothing new to be discovered in physics now. All that remains is more and more precise measurement." These were serious professional scientists, and they were not joking. But shortly thereafter, in 1900 and 1905, Planck and Einstein wrote their ground-breaking papers that led to quantum mechanics, the universe of curved spacetime, and the atomic bomb.

Compared to the late 1800s, at the present time there are a great many questions and ideas for future progress, as summarized below, so the future of science certainly does not look bleak. However, there are some practicalities that eventually cannot be avoided. Large projects in science, particularly in particle physics and astronomy, are becoming too expensive for individual countries to afford; a famous example was the Superconducting Super Collider in the USA, which would have had three times the collision energy of the LHC but was cancelled in 1993 because the projected cost had ballooned to upwards of $12 billion (in 1993 dollars!). Even with international collaborations, some future projects will become so expensive that they will never be built. And expensive projects compete with many less expensive areas of science, requiring difficult choices between different areas of science. Another limitation is the number of scientists that the population is willing to support. As the growth rate of the number of scientists considerably exceeds that of the total population, there is obviously some upper limit to the number of scientists that society would be prepared to support. So while there continue to be exciting prospects for science, the steep growth curve of the present will probably have to flatten off sometime in the future.

5.2 Will We Ever Go Backwards (Again)?

We have seen science rise and fall three times over the last two and a half thousand years, as illustrated in Fig. 2.2. On all three occasions (the Greek, Islamic and medieval epochs) the rise was rapid, followed eventually by a decline. The rapid rises can be explained by the sudden appearance of a novel and stimulating way of thinking that occurred to the early Greek philosophers in the sixth century BC and which was subsequently introduced with a sudden wave of translations into Arabic in the Islamic empire in the eight century and again with a wave of translations into Latin and the rise of universities in medieval Europe in the twelfth century. In all three cases the sharp rise was followed by a period of great activity in natural philosophy and scientific thought, and then a decline. Why did this activity disappear in the Greek and Islamic worlds, and sharply decline in medieval Europe?

The reason for the decline in Greek philosophy following its peak about 400 BC is not completely clear, and probably not simple. Greek natural philosophy was tolerated to some extent by the Roman and Byzantine empires, but was certainly not advanced by them. Perhaps it was thought that the wisdom of the ancient Greek philosophers could never be surpassed, and that everything that could be said had already been said. Christianity was a major factor in the later years. The Great Library in Alexandria was destroyed and the Platonic Academy in Athens was eventually closed by decree. Whatever the ultimate causes, the activity of Greek natural philosophy decayed away, until all that was left of it were the silent scrolls of the writings of the great philosophers.

The demise of the Islamic Golden Period of Science following its peak is much easier to explain. The religious powers became less tolerant of 'foreign' studies, and influential voices such as that of al-Ghazali railed against

Aristotelian philosophy. The madrassas threw out the Greek classics and restricted their curricula to the Qur'an. Scientific and medical books were burned by the Ulama. And the Arab world was under attack by the Crusaders and the Mongols; Baghdad was destroyed by the Mongols in 1258, along with the famed scientific academy called the House of Wisdom and its precious contents. Islamic science has never recovered.

The steep fall in scientific activity in medieval Europe in the fourteenth century was almost certainly due to the plague known as the Black Death in 1347–1350 that killed more than a third of the population. It is hard to imagine a worse disaster, causing huge turmoil and dislocation. But intellectual activity gradually recovered, leading eventually to the Renaissance and the Scientific Revolution.

One might well think that modern science will never fall away as natural philosophy and science did in those earlier periods. Modern science is quantitative and predictive in addition to explaining the world to us, and is therefore of enormous value to society. It is closely linked to technology, and underpins our modern civilization. It is now so deeply woven into the very fabric of our civilization that one would think that it could not disappear unless our entire civilization does.

But it is sobering to think just how fragile our knowledge actually is. Imagine if, by some magic, all the recorded knowledge in the world (including all books and electronic media) disappeared in one fell swoop, and all teaching stopped. Our civilization could carry on for some time, but would soon start to decay. If this went on long enough (say, fifty or a hundred years), all of the accumulated knowledge from the past millennia would be lost. Mankind would be back to a stone age existence. In just a hundred years. Of course this seems absurd, but it is not inconceivable that something like it could actually happen.

It is not completely unthinkable that some maniacal person or group gains control of the reins of power, and considers science to be an enemy. Entire libraries could be torched and intellectuals put to death. Similar things have happened before, even in recent history. The Nazis burned books and carried out the horrendous holocaust; they ravaged the scientific capability of the country, which took decades to recover. China's 'Cultural Revolution' killed tens of millions and destroyed an entire generation of academics. Cambodia's Khmer Rouge communists killed upwards of 20% of the population, many of them targeted as intellectuals because they wore glasses. Hitler, Mao Zedong and Pol Pot were apparently unrepentant to the end. Today radical Islamic fundamentalists want to destroy Western civilization and drag the world back to the seventh century. Over the course of decades or centuries it is conceivable that science could come under threat by some insane tyranny or cult, although the fact that science has now spread over the entire globe makes it less likely that an insanity in one part will be catastrophic for the rest.

How stable is our civilization? Virtually all civilizations of the past collapsed and disappeared. The Roman Empire is of course the most famous. Repeated cycles of rise and collapse were commonplace over the millennia in Mesopotamia, Egypt, India, Southeast Asia, China, Africa and the Americas. Grand structures such as Angkor Wat in Cambodia and the Mayan temples in Central America were enveloped by jungles and lost for centuries. In many cases populations fell drastically. The typical cycle lasted for hundreds of years. The causes of major collapses have been much discussed, and include climate change, tectonic events, water and soil issues, migrations, wars and invasions, depletion of resources, overpopulation, disease, cultural decline and civil war; there is no single explanation. Economic stratification can be a major destabilizing factor, and the fact that the top 0.1% of Americans have as much wealth as the bottom 90% does not auger well (the three richest alone have a combined wealth of $1 trillion—more than that of half the entire population); this stratification is sustained and enhanced by the education, control and opportunities of the wealthy and the effects of automation and offshoring of jobs on the poor. Several recent books[1] discuss the possibility that our modern global civilization, which has already lasted much longer than many others, could conceivably collapse sometime in the not too distant future if we do not manage it well.

Many books have been written about the possibility of an apocalypse of various kinds: a nuclear holocaust, a killer asteroid or comet, a monster volcanic eruption or a worldwide pandemic. An asteroid impact is thought to have wiped out the dinosaurs 65 million years ago, and the Toba super-eruption in Sumatra 75,000 years ago may have caused a global volcanic winter lasting for many years, possibly reducing the human population to just several thousand. Martin Rees has warned about various such possibilities in his book *Our Final Century* (2003), and points out that the fruits of science itself could be the killer.

Our modern Internet-based society is vulnerable, and could be severely damaged by a huge solar coronal mass ejection event such as the 'Carrington Event' of 1859 which crippled telegraph systems all over Europe and North

---

[1] Diamond (2005) Collapse: How Societies Choose to Fail or Succeed; Homer-Dixon (2006) The Upside of Down; Randers (2012) 2052: A Global Forecast for the Next Forty Years; Tainter (1990) The Collapse of Complex Societies; Motesharrei et al. (2014) Human and Nature Dynamics.

America; events of this magnitude are expected every few hundred years. But even that pales by comparison with an electromagnetic pulse (EMP) attack by some enemy. A hydrogen bomb exploding at an altitude of a few hundred kilometres would produce copious gamma-rays and an EMP that would wreak havoc over an entire continent. According to a 2017 U.S. congressional report it could shut down the electrical grid for as much as a year, decimating the infrastructure needed to support the population of hundreds of millions which can exist only because of modern technology. Electronics and electro-mechanical systems would fail. Water supplies, communications and transportation would be paralyzed. The local food supply would rapidly be exhausted, and the national supply chain disabled. Electronic payments would cease. Societal collapse could occur in weeks and mass starvation in months. Such dangers could be mitigated by hardening the critical infrastructure against such potential hazards, but the cost would not be insignificant. Similar potential threats could come from cyber-attacks by foreign powers, hacking and disrupting our Internet-based infrastructure, including power plants, the electrical grid, communications, finance and distribution networks.

A recent dramatic example of the impact of the loss of electricity was the devastation caused in September 2017 by Hurricane Maria in Puerto Rico. The entire island, with 3.4 million inhabitants, totally lost electricity (in most areas for several months), because the fragile power grid had been destroyed. The Puerto Ricans had to face up to a gloomy new and unpleasant reality. There were no lights in the homes or on the streets—it was almost totally dark at night, even in the towns and cities. A generation of young people raised with smartphones suddenly found that they couldn't communicate at all. There was no air conditioning or even fans at a time when the temperature was well above 30 °C, with high humidity and mosquitoes. Cooking wasn't possible with electric appliances. Refrigerators and freezers didn't work, and at those temperatures the food went bad in just a few days. Even supermarkets were unable to keep produce. Although the streets were flooded, there was little safe water to drink. The sewerage system wasn't working because it relies on pumps. Without working elevators and water supplies tall apartment buildings became unliveable. Gasoline pumps didn't work, so neither did cars. Electricity-based factories, offices and banks were unable to function. Schools closed. Hospitals were crippled, and had to rely on generators to produce electricity with dwindling fuel supplies. Over the course of months hundreds of those most dependent on medicines, medical devices and medical care quietly died, and others succumbed to disease; the effort to avoid a major public health disaster in a population of 3.4 million was a race against time. Thousands died.

One of the recent books on societal collapse is Dartnell's *The Knowledge* (2015), in which he envisages a catastrophic pandemic that rapidly kills almost the entire population of the world, leaving only scattered groups of tens or hundreds of survivors. Dartnell describes the scene. At first people are able to scavenge what they need to live on from the decaying contents of supermarkets and can still drive cars. But the worldwide network that kept civilization going was shattered, and it was only a matter of time before the survivors would be forced to grow their own food or gather it from trees and bushes, and hunt whatever wildlife they could find. Libraries would still exist, but most of the knowledge they contain would be of little use, unless there is a special section on survival. Dartnell gives advice on how to survive, starting by using the remnants of the lost civilization, and he gives real examples of areas in the world that have suffered severe setbacks at one time or another.

Our cherished knowledge is a 'living thing' that must constantly be nurtured and grown as new scientific and technological results become available. There is a vast industry around the world that supports and updates this knowledge in both conventional libraries and electronic media; librarians are constantly copying 'all' knowledge onto the newest storage media—it is a never-ending process. But these conventional libraries (especially electronic) may not survive a global catastrophe, and even if they did they may be of marginal use to isolated groups of desperate survivors; the esoteric details of quantum physics or molecular genetics would not be of much use in finding food for the next week. What is needed is a practical 'doomsday manual' that can be used to re-boot civilization, at least locally. There are various initiatives in this direction: Sweden recently printed such a handbook, and Dartnell (2017) proposes a portable 'apocalypse-proof' e-reader (powered by solar panels) containing the essential information.

There are other aspects of our civilization and our planet that are also protected from possible catastrophes, global or local. The Svalbard Global Seed Vault in Norway and the Millennium Seed Bank in England store duplicates of seeds held in gene banks around the world, providing insurance against the extinction of thousands of species of food crops. A Frozen Zoo in San Diego cryogenically preserves sperm, eggs and embryos for a thousand species of plants and animals from around the world. The Smithsonian's National Zoo in Washington maintains the world's largest collection of frozen milk from hundreds of species of mammals, and it is part of the Amphibian Ark, a 32-country collaboration to harbour and save amphibians from the possibility of a global die-off. The Coral Restoration Foundation in the Florida Keys maintains the world's largest collection of coral species from endangered reefs. There is an International Society for Biological and Environmental Repositories representing over a thousand biobanks around the world. A full account of global measures presently taken to mitigate the effects of possible catastrophes would undoubtedly fill many books.

5.3 What Remains to Be Discovered?

It has been quipped that "It is difficult to make predictions, especially about the future". The fact that we can now predict the motions of the planets and the properties of stars with considerable accuracy over thousands and billions of years does not mean that we can tell what will happen next year here on Earth. It is really quite sobering to think how little we can predict about the future of our lives or of human developments, including science.

Palaeofuturology is the study of past predictions about the future, and the Internet is littered with predictions that did or did not come true. Jules Verne was possibly the most successful futurologist, perhaps because he read widely and subjected his ideas to the criticisms of experts. And it may well be more difficult to anticipate new scientific developments than new technological developments. Futurologist Michio Kaku has said that "It is very dangerous to bet against the future". Lord Kelvin is probably the most famous person to have declared that "Heavier-than-air flying machines are impossible", which he did in 1895, just 8 years before being proven wrong by the Wright brothers. Another famous quote is from Einstein himself, who claimed in 1932 that "There is not the slightest indication that nuclear energy will ever be obtainable". Science fiction writers did not predict the Internet, which is possibly the greatest technical development of the past half-century.

Less well known are the 'promising fields of radio astronomy' highlighted by the prominent Australian radio astronomer Joe Pawsey in 1962, at the peak of the era of new discoveries in that field. They were ionized gas seen in absorption, magnetic fields in interstellar space, solar flares, and counts of radio sources, all fairly prosaic topics by then. Totally unforeseen were the major discoveries—quasars, the microwave background, interstellar masers and pulsars—which all occurred within the following 5 years! And again, there were the dire statements at the end of the 1800s that physics was coming to an end—just before the quantum and relativity revolutions took place. With such examples in mind, most scientists are understandably reluctant to hazard any guesses at all about future developments beyond just the next few years.

So one must tread very carefully in making extrapolations and predictions, even for the next decades let alone centuries. All one can do is to recall that the history in Chap. 2 ended with continuing questions, and to note the ongoing research, the present trends and the current technical questions in various fields of science for hints of what the near future may hold.

At the moment, fundamental physics seems to be in crisis. In spite of the enormous success of the Standard Model (SM) of particle physics, physicists are eagerly looking for any evidence of 'new physics' beyond the SM. The discovery of the Higgs boson at the Large Hadron Collider (LHC) at CERN in 2012 victoriously capped off the SM, but now, after several more years of operation, no new particle has been detected and there is no hint of anything unexpected; the SM still explains all the results. In addition, there is a major problem: the small mass of the Higgs boson. This mass, 125 GeV, agrees with years of indirect evidence, so it was by itself no surprise. But quantum mechanics predicts a value that is thousands of trillions of times bigger. The most popular proposal to reconcile the small observed value with the huge value predicted by quantum mechanics has been supersymmetry, in which every particle has a twin which contributes opposite terms to the mass of the Higgs boson, resulting in near-cancellation and a small mass for the Higgs. This fine-tuning is required for atoms (and life) to exist in our universe. But no evidence for supersymmetry or other similar proposals has been found so far in the LHC data, which remain completely explained by the SM alone.

This is reminiscent of a problem that arose in observational cosmology in 1998. By comparing supernovae in the nearby and distant universe, it was found that the rate of expansion of the universe is accelerating. The cause of this acceleration is unknown; in our ignorance it was dubbed 'dark energy', which has a repulsive force across the universe. Quantum theory tells us that 'empty space' is actually full of random 'vacuum fluctuations' which can produce just such a repulsion, except that this quantum energy is 10^{120} times greater than indicated by the observations of distant supernovae. At this enormous value no galaxies would ever have formed in the universe. As in the case of the Higgs boson, it had been hoped that this huge value may be driven to exactly zero by some as yet unknown symmetry that cancels out, but now that dark energy has been detected and is neither zero nor the huge quantum value, it poses a major problem for quantum theory. Again it appears that extraordinary fine-tuning is required for galaxies and life to exist in the universe.

In addition to these two very striking cases, there are several other 'coincidences' in the properties of our universe which also appear to be essential for the existence of stars, galaxies and life. These include the slight inhomogeneities in the very early universe, the fact that our universe has three dimensions of space and one of time, the present age of the universe, the ratio of dark matter to dark energy, the strength of gravity relative to the other forces, the ratio of the electromagnetic and strong forces, the strength of the weak force, the neutron-to-proton mass ratio, and the existence of a critical excited state in the carbon nucleus. All of these together give the impression that our universe may be fine-tuned in a way that allows for stars, galaxies and life as we know it to exist. Fred Hoyle said that it was all so unlikely that it made the universe seem like a 'put-up job', and several other scientists have said that it calls for a special explanation.

The favourite explanation is the 'anthropic principle': we can only exist in a universe that has made our existence possible. This sounds like a mere tautology, but it has been taken further.

Various independent lines of study over the last several decades have led to the idea that our universe may be just one of many universes in a vast (perhaps infinite) 'multiverse'. The concept of cosmic inflation introduced by Alan Guth in 1979 led to the idea of an infinite and eternal sea of universes that are constantly being formed at different times—inflating regions each of which ultimately becomes a separate universe. The different universes could have different laws of physics and different properties. Each of the 10^{500} possible solutions to string theory (see below) may be realized in different universes in the multiverse. And according to the 'many worlds' interpretation of quantum mechanics anything that could happen does happen in a forest of parallel universes.

It has been suggested that the multiverse concept could provide an explanation for the apparent 'coincidences' and fine-tuning mentioned above. If the different universes all have different properties and laws of physics, then a subset could exist which happen to have the conditions required for life as we know it. As we are here, our universe must be one of those. The other universes lie beyond our causal horizon, so we can never detect them. For some scientists the multiverse is an attractive hypothesis. For others it seems unscientific if there can never be experimental or observational evidence of any of the other universes (for them there is just one universe—ours—and obviously it must have the right conditions, as otherwise we would not be here). In the multiverse concept the laws of physics and properties of our universe would be random—just due to the 'weather conditions' in our local patch of the multiverse—and therefore beyond understanding. It would seem to be the end of physics.

All of this highlights the importance of the LHC and the search for evidence of 'new physics' beyond the Standard Model. As the LHC reaches higher and higher energies thousands of physicists are poring over the huge amounts of data being produced, looking anxiously for the first hints of 'new physics'. It is an exciting, if tense, time for physics.

Aside from the problems described above, what else would motivate significant changes to our model of fundamental physics? For one thing, there are various reasons to regard the SM as unsatisfactory in spite of its extraordinary successes: its many parameters and particles (one seeks the greatest possible economy in our physical theories), the probable instability of the electroweak vacuum, the masses of the neutrinos, the unnatural hierarchy of the mass scales, the inability to explain the predominance of matter over antimatter in our universe, and the inability to explain the dark matter in the universe.

One possible future step would be to unify the electroweak and strong forces into a 'Grand Unified Theory'. Supersymmetry, mentioned above, would also alleviate some of the present issues with the SM, it would provide a natural candidate for the cold dark matter, and it may play an essential role in string theory, described below.

But the big step would be to include the gravitational force itself. This is very difficult because Einstein's theory of gravity involves smooth variations in space and time, the opposite of the abrupt discreteness of quantum physics. Over the last several decades a new theoretical concept has had the attention of a large number of theorists: *string theory*, according to which the most fundamental entities are 'strings' in an 11-dimensional space. The myriad particles are just manifestations of the various vibrational patterns that strings can produce. Of great importance is the fact that string theory may possibly unify quantum mechanics and gravity into a grand 'Theory of Everything'. However it has several problems, including an excessive number of solutions and no experimental evidence, and not everyone thinks that it is going in the right direction.

Recently a new, exciting and perhaps revolutionary idea has come to the fore. Quantum entanglement, which appears to be independent of space and time, may actually be a deeper and even more fundamental aspect of reality, giving rise to spacetime itself as an emergent phenomenon. Entanglement may be necessary for space and time to exist, knitting them together into a smooth spacetime and providing a quantum theory of gravity. At the moment this is highly speculative, but hundreds of physicists are now involved and there is significant progress. These are heady days for theoretical physics.

The search for dark matter has gone on for decades, and has still come up empty-handed. XENON1T near the Gran Sasso tunnel in Italy and PandaX in Sichuan China are respectively the largest and deepest underground experimental facilities in the world, and both reported negative results in late 2017. And experiments at CERN and several other laboratories and observations by space-based telescopes have also not found any hints of dark matter. The most popular proposal, made in the 1980s, was that the dark matter may be weakly interacting massive particles called WIMPs, but this now seems doubtful, and the theorists are casting about for other possible candidates, perhaps axions (akin to massive photons). Other, far less popular, possibilities are that dark matter may exist in a 'hidden sector' and not interact with normal matter, or that dark matter does not exist at all, the dynamical effects seen in galaxies and clusters of galaxies being explained by modifications of the laws of gravity itself (unlikely given what we know about the large-scale properties of the universe, and unpopular because physicists do not want to tinker with the laws of physics). The search goes on.

The big questions in cosmology and astronomy overlap with those in fundamental physics, because the very early universe was very small, hot and dense, comprised of elementary particles and fields of force, and its possible origin (and the whole multiverse concept) is a topic in both disciplines. While the other universes in a multiverse (if there are any) would not be observable, it may be possible to find evidence for inflation in the very early universe, such as polarization in the Cosmic Microwave Background. And perhaps other clues as to the possible existence of a multiverse might eventually be found in our own universe (it has even been suggested that an unusually cool region in the CMB might be a "bruise" in our universe resulting from a collision with another universe), but at the moment this remains highly speculative.

Dark matter and dark energy are also overlapping topics for fundamental physics, cosmology and astronomy. Ever better determinations of the cosmological parameters from observations of the microwave background and the distribution of intervening matter may provide important clues. The recently-discovered gravitational waves from a binary neutron-star merger have made it possible to determine the absolute distance to the source, and therefore the expansion rate of the universe, without using the shaky cosmological 'distance ladder'. And it may soon also become possible to observe the acceleration of the expansion of the universe *directly*, using the super-large telescopes of the near future to make precise observations of absorption lines in the spectra of quasars (at present the evidence for acceleration is indirect, from observations of distant supernovae).

The evolution of the universe from the Big Bang to the present is being mapped out with increasing clarity. The Cosmic Microwave Background (CMB), which has provided such a wealth of cosmological information, is seen as it was 380,000 years after the Big Bang. Following that epoch, the universe went through a period called the cosmological dark age, when the matter was neutral and emitted no light. Gradually, the matter became more and more concentrated into dense regions in which the first stars and galaxies eventually formed and illuminated the universe again. This transition phase is called the reionization epoch, and it occurred when the universe was several hundred million years old. We can see the 'near side' of the reionization epoch from the sharp decrease in the number of quasars and galaxies as we look back from the present to that time. The reionization epoch is the last frontier of classical astronomy, and is a major target for the next generation of large ground-based optical and radio telescopes currently being developed, as well as the James Webb Space Telescope (JWST) to be launched in 2020. It should be possible to study the first stars and galaxies at near-infrared wavelengths, and the neutral hydrogen of the dark age itself at long radio wavelengths (a recently reported tentative detection using a small radio telescope, if confirmed, would be a major discovery). The irregularities in the neutral hydrogen of the dark ages, if they can be mapped, may give important cosmological information, supplementing that from the CMB.

As mentioned earlier, an important new window on the universe has just been opened with the first detection of gravitational waves in 2015—distortions in spacetime itself caused by the merger of two supermassive black holes. These black holes are so massive (about 30 times the mass of the Sun) that it is thought that they may be the remnants of some of the first stars in the universe. So the new field of gravitational wave astronomy will shed light on the astrophysics of collapsed stars, the population of massive black holes and implications for the epoch of the first stars, and perhaps even the Big Bang itself.

Another new window on the universe is neutrino astronomy. As mentioned in the previous chapter, huge new neutrino observatories are presently being completed, and exciting new results may be expected in the coming years.

A rapidly growing area of astronomy is that of extrasolar planets. Since the first discovery in 1995, there are now over 3700 known planets outside of our solar system, orbiting other stars. It is thought that the number of planets in our galaxy may exceed the number of stars, so there may be hundreds of billions of planets in our galaxy. The characteristics of these extrasolar planets are being studied in increasing detail—their orbits, their masses, even the chemical composition of their atmospheres. Of special interest are those that lie within the 'habitable zone'—the range of distance from the parent star that is 'just right' for life—not too hot and not too cold. In our own solar system only the Earth and Mars lie within the habitable zone. So the study of extrasolar planets is highly relevant to considerations of the possibility of life elsewhere in the universe. At the same time the search continues for hints of possible extraterrestrial life in our own solar system, perhaps on Mars or the moons of the giant planets. The big question is "Are we alone in the universe?"

In our solar system the most likely bodies on which extraterrestrial life might be found are Mars and some of the moons of Jupiter and Saturn. This life would probably be microscopic (we still don't know exactly how to define life), but its discovery would nevertheless be of monumental importance. One fundamental question would be whether it has the same chirality ('left-handedness' or 'right-handedness') as life on Earth. All life forms on Earth have the same chiralities: DNA is always right-handed and amino acids are always left-handed. If the extraterrestrial life is different in this regard, this would prove that it had formed independently from that on Earth, making it all the more likely that life has formed throughout the universe—that the universe is teeming with life.

One of the very big questions in biology overlaps with astronomy—the origin of life. In astronomy, not only do we study the rapidly proliferating exoplanets themselves, we can also study the actual processes of the formation of planets. Giant radio telescopes, in particular the Atacama Large Millimetre/Submillimetre Array (ALMA), can peer deep into the dust-enshrouded cores of star formation regions with high resolution and sensitivity. Gaps caused by forming planets can be clearly seen in images of the protoplanetary disks, and spectroscopy reveals the thousands of spectral lines from organic molecules—possible seeds for future life on those newly forming planets. ALMA has only been in operation since 2013, and we can expect quite a number of important discoveries in the coming years.

How life on the early Earth emerged from abiotic chemistry remains a mystery. Fossil records of life that existed 3.5 billion years ago have been found, and the Earth itself formed 4.6 billion years ago, so sometime in the intervening billion years life was somehow able to form. Laboratory experiments have been recreating the likely conditions at that time; they have succeeded in creating amino acids—some of the building blocks of life—but not life itself. Detailed studies of possible ways that prebiotic molecules could have come together to form living systems are being made, but it is a complicated matter and progress is slow. An alternative approach is to try to make life ourselves—'artificial life'. Then the actual way that life originally formed on the early Earth would just be a matter of history, rather than of fundamental science. It has been possible to create synthetic DNA from 'off the shelf' chemicals, and ways in which cell membranes can 'self-assemble' have been studied, but these steps, important as they are, are still a very long way from producing a complete living system from scratch. Nevertheless, the optimists in the field think that true artificial life may eventually be created in the laboratory.

The big question in cell biology hasn't changed for many decades: what is between the genome and the phenotype (the observed characteristics of an organism), i.e. what is the 'epigenotype'? It is a huge question, really many questions in one. There is an enormous number of combinations and networks of interactions and layers that communicate information from the genome to the phenotype. Some of the discoveries over the past several decades were mentioned in Chap. 2. Unlike the case of sequencing of the human genome, which had a fixed end-point, we do not have a clear idea of the possible outcomes of many of the current lines of research. As we learn more, we uncover still more complexities. To cure cancer and other diseases we first have to understand the epigenotype, and that may take quite some time. Such diseases are ultimately caused by the underlying fundamental process of ageing itself, which is slowly beginning to be understood. Progress on understanding ageing will help with everything else, and may someday make it possible to actually reverse the effects of ageing, at least for some of the 200 tissue types in our bodies.

The explosive development of biology has resulted in a wide variety of new topics now being explored. One fascinating question is how we humans evolved to be so different from other animals. The number of genes in our genome is far less than that of wheat and our brains are only three times the size of those of chimpanzees. So how to explain our vastly superior brainpower? It is obviously not the size of the genome or the brain. It appears most likely to be due to greatly enhanced complexity in interactions and modulation of gene expression. The availability of complete sequences of the genomes of modern humans, archaic humans, chimpanzees and several other animals and organisms has opened up the whole field of differential evolution. It is found that there are DNA sequences that are conserved in other animals but have changed rapidly in humans; these are called human accelerated regions (HARs). There are thousands of them; they inhabit the regions of the genome previously called 'junk DNA', and almost all of them are 'enhancers' that modulate the expression of genes. So perhaps the evolution of our huge brainpower may have been due to the HARs and their ability to create systems of great complexity. They were discovered just 10 years ago, and this exciting field is rapidly developing.

New tools are being developed almost as fast as our understanding of molecular biology. CRISPR was named the breakthrough of the year in 2015 by the American Association for the Advancement of Science, and there is now talk of the possibility of 'precision medicine' in future years. One of the big breakthrough applications of CRISPR is the ability to tag and trace the history of every cell in an organism. CRISPR has also recently been used to create a synthetic cell that contains the smallest genome of any known, independent organism, functioning and reproducing with just 473 genes. The ultimate goal is to build the smallest possible genome from scratch, so the function of every gene can be understood; then more complex living systems can be engineered and built with certainty about their functions.

A completely different kind of technical advance makes it possible to examine and stimulate individual neurons in the brain, and in another advance scientists have recently been able to precisely measure the activity of hundreds of neurons at a time, opening up a whole new field—the study of functional neuronal activity in real time. New discoveries continue to be announced, and it's a very exciting time in all of these fields.

But the big theme in biology remains "How do systems work?" This is the source of almost all of the big questions in biology, including the epigenotype, and it is a huge challenge. How does metabolism as a system work, how does it interface with the environment and with the genome, how does it enable natural selection to affect change, how is it related to disease? Various strategies have been tried, but they have fallen far short of the major long-term objective, which is to understand the fundamental signatures in complex systems of all kinds. More and more tools are being developed and knowledge advances, but the problems remain daunting. There is much to be done and various approaches are being taken, and we have no idea what the discoveries, twists and turns in this field will be in the future.

So, in all three of the areas considered above—fundamental physics, astronomy and cosmology, and biology—profound and deep mysteries remain, and we can look forward to exciting discoveries and scientific advances over the coming years and decades.

But can we make an educated guess as to the longer-term development of science? Martin Harwit tried to do this for the relatively well-defined field of astronomy in his 1981 book *Cosmic Discovery*, as discussed in Chap. 3. Astronomy is particularly amenable, as it relies on pure observations using a small number of observational windows (experimental science would be much harder, because the conditions and parameters are endlessly variable). Harwit considered a 'multi-dimensional observational parameter space' in which a 'discovery' is a phenomenon that is separated from others by at least a factor of a thousand in any one of the parameters. Harwit gives a number of examples to show that this criterion does indeed distinguish between different known phenomena.

Harwit notes that such phenomena have sometimes been discovered twice in two completely independent ways using two separate instruments that differ by a factor of at least a thousand in one of their observing capabilities. He suggests that such duplications might indicate that the number of potential discoveries in the multidimensional parameter space is finite, and that the discovery of more duplications (and triplications) may give us a way of estimating the total number of potential discoveries in the entire multidimensional parameter space—an estimate of the total number of cosmic phenomena that can be discovered. Using this ingenious approach and the data he had available at that time, Harwit estimated that the total number of potential discoveries is about 130, and that we had already found one-third of them by that time (1981). This is of course only a crude estimate, and there may certainly be critics, but the concept is intriguing.

5.4 Can Science Ever Be Complete?

Are there any limits to science? Can we eventually know everything?

In the years following the publication of Newton's *Principia* one might have thought that we could know everything about the world, at least in principle. It was a 'clockwork' universe, and all events—past, present and future—were connected by strict causality, and could be known with unlimited accuracy.

However, we now find that we are confronted with a number of roadblocks and obstacles, and the ideal of complete knowledge has receded from our reach.

One obvious issue, as we rely on observations and experiments to know the real world, is the ever-present problem of measurement uncertainty. We can always reduce the uncertainty by increasing the duration of the observations or the number of experiments, but some uncertainty, however small, will always be present.

Another issue is the overwhelming scale of the real world—the vast number of particles in the universe, and the immense complexity of living systems. The first of these can be dealt with by understanding the governing principles rather than recording individual cases, and physics has been quite successful in this endeavour. Physicists dream of a 'Theory of Everything' (a 'Final Theory'), which is a complete theory of fundamental physics. It would contain all the physics that underlies everything in the natural and physical world. The current version of the 'Equation of Everything' is shown below (Eq. 5.1). *In principle* a Theory of Everything could unfold to explain and predict everything in the universe, but in practice there is a huge gap between the elegant simplicity of physics at the most fundamental level and the vast complexities of the world around us.

$$\Psi = \int e^{\frac{i}{\hbar}\int\left(\frac{R}{16\pi G}-\frac{1}{4}F^2+\bar{\psi}i\not{D}\psi-\lambda\varphi\bar{\psi}\psi+|D\varphi|^2-V(\varphi)\right)}$$

Schrödinger Feynman Einstein Maxwell-Yang-Mills Kobayashi-Maskawa Euler Planck Newton Dirac Yukawa Higgs

Equation 5.1 Equation summarizing all the known laws of physics (Turok 2013). It includes gravity, the three forces of particle physics, all the matter particles and the Higgs field. The symbol on the left (ψ) represents Schrödinger's wavefunction, the most complete description of a physical system

Biologists are absolutely shocked at the idea of anything like a 'Theory of Everything'. They find it absurd to even contemplate such a concept; they are certain that science can never be complete. Their reaction is understandable, as they confront the vast complexity of living systems on a daily basis. They deal with the most extreme complexities that exist in the universe, in contrast to the physicists, who work towards the ultimate simplicity of the underlying world (science is relatively simple on both the largest and smallest scales, but biology sits in the middle where the complexity is greatest). Evolution never stops, so they are dealing with ever-changing systems and an entire biosphere with vast numbers of overlapping and interleaving entities, activities and influences. It will be a monumental task to understand the fundamental principles of these systems. And even with these principles, biologists think that it may never be possible to make accurate predictions. Even simple and deterministic first-order difference equations can lead to apparently random and bizarre outcomes.

What about other areas of complexity? We're doing a fairly good job with the weather on time scales up to a week or two, but will we ever be able to predict the 'butterfly effect' (e.g. storms in the northern hemisphere caused by the flapping of a butterfly's wings in the Amazon jungle)? Even with reliable physics how would we ever be able to know the 'initial conditions' with any precision? Will we ever be able to predict earthquakes reliably? How much will we ever learn about the interior of the Earth?

Almost all of our observations of the universe are made by detecting electromagnetic radiation, and there are some practical limits. At very low (radio) frequencies, our view becomes obscured: the Earth's ionosphere absorbs the radiation below 1 MHz, and the interstellar medium absorbs the radiation below about 100 kHz. And at the high frequency end of the spectrum the gamma-rays interact with the photons of the Cosmic Microwave

Background, obscuring galaxies beyond our own. Similar limitations apply to another window on the universe, cosmic rays; at low energies the solar wind blows them about and into the interstellar medium, and at high energies they are destroyed by interactions with the photons of the CMB, limiting their use to the nearest galaxies. Furthermore, cosmic rays, being charged, are deflected multiple times by the complex interstellar magnetic fields, so the locations of their sources are difficult or impossible to determine. In principle we could learn more if we were to someday undertake extensive space travel, but even then our knowledge would be plagued by limitations of these kinds.

The speed of light is finite, and nothing—not even information—can travel faster than the speed of light *in vacuo*. Because of this we can't see the expanding universe beyond our 'light cone'—our view of the universe is limited (and more so with increasing time, as the expansion is accelerating). It remains unclear whether any infalling information could ever be retrieved from a black hole. We can't see all the way back to the Big Bang (using electromagnetic waves) because of the obscuring Cosmic Microwave Background. Even if the multiverse scenario described in an earlier section turned out to be true, we would never be able to detect the other universes. Furthermore, the laws of physics in our own universe would then probably be due to random 'weather' in our local patch of the multiverse—they would not be fundamental.

In our search for the most fundamental constituents of matter we have built giant accelerators that cause particles to collide with one another at speeds close to that of light. The chaos produced by these high-energy collisions is examined for signs of ever more fundamental physics. But we are rapidly approaching the point of diminishing returns, and may soon not be able to afford the huge cost of larger and larger accelerators required to reach the highest energies. Another potential limitation in our quest to understand fundamental physics could be variations in the laws of physics themselves. At the moment we have very strong upper limits on the variations of certain fundamental 'constants' (less than a million billionth per year), but even tiny variations in time or space could limit our ultimate understanding of physics.

Heisenberg's uncertainty principle, mentioned in Chap. 2, imposes a fundamental limitation on our knowledge of the subatomic world. The position and the momentum of a particle cannot both be known with unlimited precision; if the position is accurately known, the momentum cannot be, and vice versa. The same applies to time and energy. This uncertainty is intrinsic in nature—it is not a result of measurement error. It is a central feature of quantum mechanics, and cannot be avoided. The subatomic world is determined by probabilities.

There are other limitations on our ability to have complete knowledge. As Karl Popper stressed, a theory can never be proven to be true; it can only be falsified. No matter how many experiments have been made that agree with the theory, they do not prove that the theory is correct. But it only takes one experiment that disagrees with the theory to prove that it is false. So we can never be sure that our theories of the world are absolutely correct.

Another limitation on our theoretical models of the subatomic world is that they may be non-unique. While we may have one model that fits all the data, we may never be sure that there couldn't be a different model that equally well fits the data. An example was mentioned above in Chap. 2. In the early days of quantum theory, Erwin Schrödinger gave a mathematical model describing the behaviour of electrons in atoms in terms of probability waves, and Werner Heisenberg described the same phenomena in terms of quantum-jumping between energy levels. In one case a wave equation, in the other case matrix mechanics. Paul Dirac produced a more abstract formalism, and showed that all were mathematically equivalent. And, more generally, wave-particle duality is a common theme in quantum mechanics. The best we can do is make models that *work*, in that they reproduce our experimental and observational results.

And, of course, any infinities in space, time or density could render our scientific knowledge incomplete.

Even if we were able to produce a 'Theory of Everything' that is unique and explains all observations of the world on all scales, we would never know for sure whether it encompasses absolutely everything that exists. There could always be something more that eludes us. And even if such a theory did in fact contain everything that exists, while science would then be complete we would never be able to prove it.

Are we missing anything at a still more profound level? Is there some completely different way of seeing the world that is all around us and right in front of our eyes, but of which we are still totally unaware? Science covers everything that exists in the natural and physical world, but does it do so in all possible ways? Many physicists would say that we still lack a deep understanding of quantum mechanics, in spite of its phenomenal success. Are we in the same position as the ancient Greeks before they thought of natural causes?

At a somewhat less fundamental level this is reminiscent of our situation in the nineteenth century before Maxwell's theory of electromagnetism. We knew only about the tiny optical part of the spectrum, which covers less than a trillion trillionth of the wavelength range of the whole electromagnetic spectrum; we had absolutely no awareness of the existence of the radio, infrared, ultraviolet, X-ray and gamma-ray bands, all of which we use today. So, what are we missing now?

One last and even more esoteric limitation: if, in spite of all the above limitations, we somehow achieved complete knowledge and wanted to encode it in a set of axioms from which all knowledge could be deduced, there would be a limitation imposed by Gödel's incompleteness theorems which prove that no significant system of mathematics can be produced which is both consistent and complete. Science would be limited by the incompleteness inherent in axiomatic mathematics.

So, on the basis of our present knowledge, it seems that there are many hurdles preventing us from ever achieving the goal of absolute completeness in science. Perhaps developments in the science of the future will change this picture, but at the moment it seems unlikely. We live in the real world, not the Platonic world of perfect ideals.

However, it is not inconceivable that science may someday in the distant future be reasonably complete in a more pragmatic sense, providing a 'working knowledge' of all the things we can detect. We have to accept that our measurements can never be perfect—there will always be uncertainties. We also have to accept that there are practical limitations to the power of our observational and experimental facilities. We may still be overwhelmed by the complexities of life. There is no way of avoiding the limitations imposed by the uncertainty principle. Theories, no matter how well confirmed, will always in principle remain provisional, in that there is always the possibility that some future measurements may disagree with them, forcing modification or even radical change. We will have to live with the fact that our theoretical concepts and models of the subatomic world, however successful, may be non-unique; they are after all just conceptual tools that enable us to make predictions that work. This is the real world, and we do our best to understand what is in it and how it all works.

We've already come a long way in that direction, with Newtonian and Einsteinian physics which apply both on the Earth and throughout the universe, quantum mechanics that underpins all of chemistry and the electronics revolution, the Standard Model of particle physics that explains all of the experimental results from the world's particle accelerators, and an understanding of the evolution of the universe, the evolution of life, and the genetic basis of life. Who knows how much further this road will take us?

5.5 The Long View

In this book we've only been talking about hundreds or thousands of years.

That's nothing compared to the *billions* of years of the universe. The beginning of the universe was 13.8 billion years ago (bya), the formation of the Sun and Earth occurred 4.6 bya, life emerged on Earth by 3.5 bya, complex life arose about 1 bya, and our speciation from the chimpanzees took place 7 million years ago. Recorded history only goes back 5000 years, the Scientific Revolution was just a few hundred years ago, and we've only had modern technology for the past hundred years or so—a tiny fraction of our existence as a species. Where are we headed?

At present we have a long list of scientific questions in front of us as we look into the future. These questions will keep us busy for years to come, and many unexpected discoveries in the future will lead to further questions. Exciting times for science.

So exciting that one may wonder whether, without realizing it, we are actually presently living in the 'Golden Age of Science', when most of the important and fundamental discoveries that can ever be made are currently being made.

On a timescale of billions of years, this Golden Age of Science may be a mere blip—a very exceptional time in our existence as a species.

That would of course imply that eventually there will be a time—a hundred, a thousand, a million years from now—when most of the important and fundamental questions will have been answered. Science would undoubtedly continue, perhaps dealing increasingly with questions of less and less importance, and of course our curiosity will never die, but the bulk of the scientific activity would be over.

If there is such a time in the future, we would then look back wistfully at the past Golden Age of Science in which we presently have the good fortune to live. We would be entering the 'post-discovery world' (Fig. 5.1).

Will science by that time have made us very different from what we are at present? In spite of all the scientific and technological progress over the last few hundred years, we haven't been *totally* transformed.

But actually, over the last several thousand years, we *have* significantly changed both ourselves and the world around us. Ten thousand years ago we still lived as hunter-gatherers, as we did over the previous several million years; 99.9% of our existence was spent in that state. But things began to change, slowly but significantly, after the advent of agriculture. Over time we caused dramatic evolution in several species through selective breeding. We

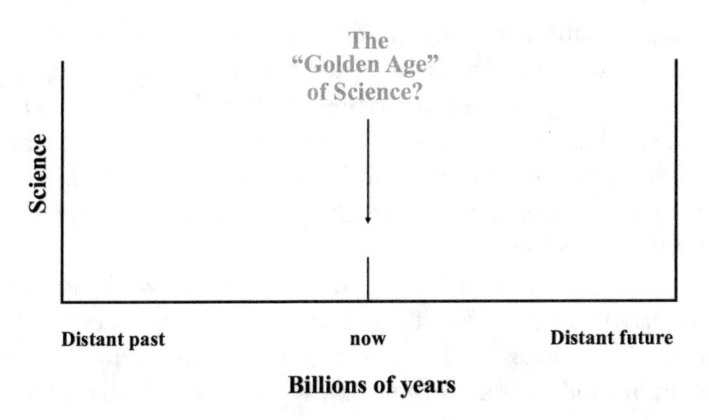

Fig. 5.1 Science as a function of time, on a scale of billions of years. If the rate of scientific discovery declines hundreds, thousands or even millions of years from now, the present 'Golden Age of Science' would be a mere blip in this plot

have changed the face of the planet itself, with vast areas cleared for cultivation or for grazing by our domesticated animals, and landscapes criss-crossed with roads and railways, and dotted with towns and cities.

Our own evolution may also have accelerated over the years, and may now be as much as a hundred times faster than it was a few million years ago according to some studies. One factor may be our much greater population, which can produce far more beneficial mutations capable of spreading more rapidly. Another may be innovation and competition in our increasingly technological world. In recent years our lives have changed dramatically in many ways. We live in temperature-controlled conditions, we have reliable and wide-ranging sources of food, and we have protected ourselves from many diseases. We have artificial aids for hearing and seeing, pacemakers, wonder drugs, transplants, artificial body parts and we are starting to make artificial organs. There have been huge medical advances at both ends of our lifespan. Our life expectancy has more than doubled over the past century. Our lives are changing rapidly.

And genetics may become a major game-changer. Astonishing progress is being made on a number of fronts. The revolutionary new technology called CRISPR makes it possible to cut, paste and edit any DNA sequence into or out of a genome quickly and easily. The applications are endless. Bacteria and other life forms with new traits can be engineered in just weeks. The most important and controversial application may ultimately be the modification and evolution of humans, but for this much discussion and societal approval would obviously be required. 'Synthetic cells' have already been made in which an organism is controlled by artificially designed and programmable DNA;

potential applications include the production of fuels, oils, proteins, vaccines, materials and antibiotics. And in 2014 the first living and propagating organism with an artificial genetic code was produced; the bacterium's code contains six letters instead of the usual four. The fundamental code of life was changed and a completely new evolutionary tree created, which can live in parallel with but fundamentally different from all existing life forms on Earth, resistant to all other bacteria and viruses.

So we have advanced from being part of the natural world and completely subject to Darwinian evolution, to controlling the resources of the planet and the lives of other animals, and to now being able to contemplate controlling our own destiny and creating entirely new types of life. These are huge steps in the history of humanity.

Where will this take us next? Some in the field have commented "now that we know we can rewrite life, we are very unlikely to stop".[2] We've only been working on the genome for the last several decades, and the cell is far more complex than DNA itself. Where will we be in a thousand or a million years? Will we in that future time have changed our chemistry, escaped from natural evolution and be in total in control of our own destiny? Even if possible, would that be wise? We have benefitted enormously from 3.5 billion years of natural evolution. It has given us impeccably finely-tuned defences against all sorts of potential ills, germs and disorders. Wouldn't we be unwise to abandon all that and try to take control of our own evolution? Possibly. But we have been responsible for the past ten thousand years of evolution of the grey wolf into all sorts of dogs, and that certainly hasn't killed off the dogs, although that was (just) done by selective breeding. It is the idea of meddling with the human genome directly that rings the most alarm bells. On the other hand, we have decades, centuries or millennia to do this if we choose, taking great care along the way and learning much more than we presently know about the intricacies of the epigenotype. So it is at least conceivable that we could eventually safely 'evolve ourselves'. But, even then, can we trust ourselves? Some of the potential perils were mentioned in the last chapter.

If we were eventually to take that step, our own evolution (and that of any other species we choose) could become enormously faster than Darwinian evolution by natural selection, as we determine for ourselves what we want to become. Would we have the potential for immortality? Would we have created abundant new forms of biological life? Would life in that future time be something we cannot even imagine now?

[2] Enriquez and Gullans (2015) Evolving Ourselves: How Unnatural Selection and Nonrandom Mutation Are Changing Life on Earth.

But even such developments in genetics may pale in comparison with those of 'artificial intelligence' (AI).

The idea of advanced computers being capable of thinking for themselves (and being conscious) has been around for over half a century, but in spite of much hype it was hopelessly far ahead of its time. We have had to meticulously programme the computers ourselves—they could only do exactly what we told them to do. But recently, not only have they become extremely powerful (billions of times faster than the human brain), they are also becoming capable of learning. They are mastering an increasing number of tasks, such as playing chess and Go, and text, speech and face recognition; they are better and safer than humans in flying and landing aircraft, and autonomous vehicles are on the horizon. Long-term objectives include planning, reasoning and complex robotics. Huge investments are going into AI, as it may prove to be revolutionary. At the same time much research is underway to develop 'quantum computing', which may replace the present digital computers with 'analogue' quantum computers of vastly greater power. AI is at the cutting edge of technology—so much so that it has been quipped that "AI is whatever hasn't been done yet".

Could sufficiently powerful and sophisticated computers and AI someday in the distant future render human brainpower (and us) superfluous? Various studies indicate that there are physical and metabolic limits to the potential power of organic brains such as ours. And the information saved in our brains from years of education and experience is dispersed and lost along with the atoms of our bodies when we inevitably die; because the 'carrier' is mortal every new generation has to learn everything all over again. There are no such limitations for computers, which could be immensely powerful and immortal. The total information content of our civilization, which now far exceeds the capacity of any single human brain, could be readily accessible to highly advanced 'thinking machines'. It would never be lost, and could be shared amongst a vast number of them. It is also conceivable that the thinking machines of the future may be capable of studying and learning about the world as we do now. And it is not completely out of the question that they could develop self-awareness, consciousness and the ability to plan and create as we do now but on a far grander scale. All of this is just wild speculation at the moment, but—as computers have only been around for less than a century and here we're talking about possibilities for the next thousands or millions of years—might it someday be possible?

In the very long term, might our present era of human biology and brainpower ('wetware') be just a brief blip in the history of intelligence in the universe? Might our creations—thinking machines—take over and continue the advance of progress that we initially set in motion? If it is our heritage that they take forward, then they would not really be 'taking over'—they would be truly our descendants. It may ultimately be our robust thinking machines, rather than fragile biological humans, that colonize the galaxy.

* * *

Science fiction is replete with extrapolations from modern science for interstellar travel employing wormholes and other similarly extravagant ideas. Which (if any) of these will become reality? Although the Apollo programme 50 years ago was inspiring, there has been a notable hiatus, and human travel to the distant stars remains a remote dream. It seems that we'll be stuck here in our solar system for a long time to come.

However, the spacecraft Pioneer 10, launched in 1972, has gone beyond the heliosphere and will eventually leave the solar system entirely, and serious thought is now being given to sending tiny spacecraft to Alpha Centauri, the nearest star system to the Earth (just 4 light-years away), using ground-based lasers to propel them to 20% of the speed of light so they can make the trip in 20 years. Other concepts for long-distance space travel have long been pondered, including ion and anti-matter propulsion systems. Plans to eventually colonize Mars (9 months' travelling time from Earth) may someday come to fruition, and would provide extra insurance for the survival of our species; we currently have a 'window of opportunity' for this—while we still have the required technology and before we destroy the Earth. But these are tiny baby steps compared to, for example, colonization of the entire galaxy. It has been estimated that, *if* we could master all the required technologies for propulsion, survival and procreation, the galaxy could in principle be colonized in as little as 10 million years. As this is very short on cosmological time scales, it raises Fermi's famous question "Where is everyone?"—why haven't our galactic neighbours already been here?

The Big Bang occurred 13.8 billion years ago, and the first stars and galaxies were formed over 13 billion years ago. A huge number of stars will have formed over the billions of years preceding the formation of our Sun and Earth 4.6 billion years ago. If life is common throughout the universe, it could have arisen a great many times, billions of years before life arose on Earth. Those early life forms would now be billions of years more advanced than we are. If so, we are total newcomers to the communities of the universe. We may be in the position of present-day jungle dwellers who communicate with drums, completely unaware that there exists a vast global network of radio communication whose waves pass right through them.

What do we know about the possibility of life elsewhere in the universe? To make it simple (as we have no idea what exotic forms 'life' could conceivably take in the universe), we restrict our quest to life as we know it here on Earth, and start by considering what Earth-like planets there may be elsewhere in the galaxy and the universe. The first planets outside our solar system were found in the 1990s. Since then, from intensive searches made over the past two decades, we have found over 3700 of these 'extrasolar planets', and we now know enough about their characteristics and those of their parent stars to be able to estimate their demographics not only in our Galaxy but also in the entire observable universe. There are about 100 billion Earth-like planets in our Galaxy, and billions of trillions in the observable universe. The average age of those in our galaxy and others nearby is 7–8 billion years. That's a few billion years more than the age of the Earth.

The question then becomes how many of these planets might host life. A major criterion for a planet to be able to sustain life as we know it is that it should be located in the 'habitable zone' of its parent star (not too close and not too far from the star to be able to maintain liquid water). The fraction of Earth-like planets that satisfy this criterion is about 10%. Various other factors may play a role, such as interstellar clouds, stellar interactions, stellar activity, the influence of other planets in the system, orbital migration, resonances, moons and tidal interactions, asteroid and comet impacts and major volcanic events.

Another, more sinister, possibility is 'sterilization' of the planet by highly energetic radiation (gamma-rays) from the death throes of massive stars (supernovae). The gamma-ray bursts would produce chemical reactions in the atmosphere of a planet, destroying its ozone layer and thus exposing life on the surface and shallow waters to the deadly ultraviolet radiation from the parent star for years, and creating a smog which could cause a 'cosmic winter'. All life could be wiped out, except for some extremophiles and subsurface and deep-sea life.

This gamma-ray catastrophe, occurring in any regions of the universe with a high density of massive stars, could sterilize those regions. This would include the dense central regions of normal galaxies and the dense star-forming regions in the spiral arms of galaxies like ours. It would also include the relatively small star-forming galaxies that existed in the early universe; the entire universe before about 5 billion years ago may have been 'sterilized'. Other such radiation hazards include the supernovae themselves, cosmic rays, and the jets emanating from quasars and active galactic nuclei.

Our Sun with its planetary entourage happens to be located in a particularly benign region in our galaxy—between the spiral arms, and at an intermediate distance from the centre of the galaxy. Nevertheless there are indications that even our Earth may have been affected by gamma-ray events in the past. There is a high probability that Earth experienced at least one lethal gamma-ray burst over its lifetime, and a 50% chance of one in the past 500 million years; indeed the late Ordovician mass extinction that wiped out about 80% of the species on Earth 443 million years ago may actually have been caused by a gamma-ray event, and there are tell-tale signs in deep Antarctic ice cores of gamma-ray activity at the times of two known supernovae (those of 1006 and 1054). So we are lucky to have made it to where we are (but of course if we weren't we wouldn't be here to comment on it).

But on the other hand there is the sheer overwhelming number of planets in the universe, even in regions located far from the radiation hazards mentioned above. And if life is inevitable when conditions are right, it will seize the opportunity as soon as it can. Furthermore, life as we know it does seem to have a tenacious ability to cling on in the face of adversity.

So—is the universe teeming with life in spite of all the hazards? Or are we alone? What are the odds of *intelligent* life having evolved, as it has here on Earth? Are we amongst the first intelligent life forms in the universe, or are there hordes of others that are already billions of years more advanced?

Over the past half century we have been making searches for possible signals from any extraterrestrial intelligence that may be 'out there'. When the first pulsars, which are as precise as an atomic clock, were discovered in 1967, an obvious first thought was that they could be the beacons of some vast galactic civilization, and they were nicknamed LGM for 'little green men'. But it was quickly realized that they are just garden-variety neutron stars. Similarly, the mysterious Fast Radio Bursts (FRBs) that have been detected over recent years probably have natural explanations in terms of merging neutron stars or black holes, but here again some speculate that they may be due to extragalactic civilizations. Today searches continue for possible evidence of extraterrestrial intelligence in several windows of the electromagnetic spectrum.

Needless to say, if we were ever to encounter advanced extraterrestrial life, it would undoubtedly be by far the greatest intellectual, cultural, emotional and religious shock we could ever imagine.

If we have trouble imagining what our lives may be like mere hundreds or thousands of years from now, think how different the typical occupants of the universe may be—billions of years more advanced than we are. They had plenty of time to colonize their galaxies. Why haven't our galactic neighbours been here? Or have they? Is it possible that their inevitable fate was to kill themselves off, perhaps by nuclear holocaust or viral pandemic? That seems unlikely, as they should have colonized other worlds so the disasters on one would not affect the others. So are still out there, in forms we can never imagine? We currently think of intelligent water-based *biological* life on other planets, but life on other planets may be very different from us, and the 'thinking machines' mentioned above may prefer interstellar space or the environments of black holes as habitats. We have had modern technology for only the last hundred years—how can we possibly have any inkling of what the other occupants of the universe might be like—and therefore what *we* may be like billions of years from now.

The biggest developments of the far future may well come from discoveries that have not yet been made, and that we at present cannot even imagine. The mind boggles at what the future may hold for us. It will be quite an adventure.

Epilogue

Advanced science and technology have totally transformed our lives over just the last few lifetimes—a miniscule fraction of the millions of years since our early ancestors emerged.

The seeds for modern science were put in place 2600 years ago by the ancient Greek natural philosophers. The key was rejecting the religious worldview of gods and myths to explain the world, and replacing it with a scientific and rational worldview in which the causes of events are part of the natural world itself.

This 'Greek Miracle' lasted for a thousand years. It was absolutely unique; nothing like it has ever happened anywhere else in the world. It was the first fundamental step in the history of science.

Fortunately, it was recorded in precious scrolls that were eventually translated into other languages. The unique wisdom of the Greek philosophers was passed down in a thin thread through history, leading to the Scientific Revolution in seventeenth-century Europe—the second fundamental step in the history of science.

The Scientific revolution added the concept of laws of nature, quantitative predictions, and the scientific method for testing these predictions with reality. It was the birth of modern science.

From that time on science advanced exponentially, and we have now explored the atom, the universe and the basis of live itself. Advanced science merged with technology in the nineteenth century to produce the high standard of living that we enjoy today.

P. Shaver, *The Rise of Science*, https://doi.org/10.1007/978-3-319-91812-9

Science is now woven into the very fabric of our civilization. We have entered a totally new phase of human existence. Life without our scientific and technological marvels would be unimaginable. The ancient Greek philosophers would be absolutely stunned if they could see what their idea of natural causes has led to.

Science continues to grow exponentially, and we may now wonder whether this can continue, and where it will lead. Will we eventually have discovered almost everything there is to know? Science has given us great power; how we use it in the future may be our greatest challenge. Will we transform ourselves biologically? Will our own creations—computers—someday take over from us? Will we discover extraterrestrial life?

The universe is 13.8 billion years old, and may contain billions of trillions of planets like the Earth. Our existence only goes back millions of years, and we have only had modern technology for the last century. We are total newcomers to the universe. Will we encounter galactic neighbours who are billions of years more advanced than we are? And is that what *we* will be like billions of years from now?

Further Reading

Agar J (2012) *Science in the Twentieth Century and Beyond*. Polity Press, Cambridge, UK

Al-Khalili J (2012) *Black Holes, Wormholes, and Time Machines*. CRC Press, Boca Raton, Florida, USA

Al-Khalili J (2012) *Pathfinders: The Golden Age of Arabic Science*. Penguin Books, London, UK

Al-Khalili J (2012) *Paradox: The Nine Greatest Enigmas in Physics*. Broadway Books, New York

Al-Khalili J ed. (2016) *Aliens: The World's Leading Scientists on the Search for Extra-terrestrial Life*. Picador, New York

Al-Khalili J ed. (2017) *What's Next? Even Scientists Can't Predict the Future – or Can They?* Profile Books, London

Alper M (2006) *The God Part of the Brain: A Scientific Interpretation of Human Spirituality and God*. Sourcebooks Inc., Naperville, Illinois, USA

Annas J (2000) *Ancient Philosophy: A Very Short Introduction*. Oxford Univ. Press, Oxford, UK

Annas J (2003) *Plato: A Very Short Introduction*. Oxford Univ. Press, Oxford, UK

Atkins P (2011) *On Being: A Scientists' Exploration of the Great Questions of Existence*. Oxford Univ. Press, Oxford, UK

Baggott JM (2011) *The Quantum Story: A History in 40 Moments*. Oxford Univ. Press, Oxford

Baggott JM (2015) *Origins: The Scientific Story of Creation*. Oxford University Press, Oxford

Balchin J (2014) *Quantum Leaps: 100 Scientists who changed the World*. Arcturus Publ. Co., London

P. Shaver, *The Rise of Science*, https://doi.org/10.1007/978-3-319-91812-9

Ball P (2013) *Curiosity: How Science Became Interested in Everything*. Vintage Books, London, UK

Barnes J (2000) *Aristotle: A Very Short Introduction*. Oxford Univ. Press, Oxford, UK

Barrat J (2013) *Our Final Invention: Artificial Intelligence and the End of the Human Era*. Thomas Dunn Books, New York

Bennett J, Shostak S (2007) *Life in the Universe*. Pearson/Addison-Wesley, San Francisco, California, USA

Bering J (2011) *The Belief Instinct: The Psychology of Souls, Destiny, and the Meaning of Life*. W.W. Norton & Co., New York

Bertman S (2010) *The Genesis of Science: The Story of Greek Imagination*. Prometheus Books, Amherst, New York

Beyret et al. (2018) *Elixir of Life: Thwarting Aging with Regenerative Reprogramming*. Circulation Research 122, 128-141

Bickerton D (2009) *Adam's Tongue: How Humans Made Language, How Language made Humans*. Hill and Wang, New York

Bignami GF (2012) *We are the Martians: Connecting Cosmology with Biology*. Springer, Heidelberg

Bignami GF (2014) *Imminent Science: What Remains to be Discovered*. Springer, Heidelberg

Börner G (2011) *The Wondrous Universe: Creation without Creator?* Springer, Heidelberg

Bonnet R-M, Woltjer L (2008) *Surviving 1,000 Centuries: Can we do it?* Springer-Praxis Publ., Chichester, UK

Bornmann L, Mutz R (2014) *Growth Rates of Modern Science: A Bibliometric Analysis Based on the Number of Publications and Cited References*. Journal of the Association for Information Science and Technology 66.10.1002

Bowler PJ, Morus IR (2005) *Making Modern Science: A Historical Survey*. Univ. of Chicago Press, Chicago

Brockman J ed. (2006) *What We Believe but Cannot Prove: Today's Leading Thinkers on Science in the Age of Certainty*. Harper Perennial, New York

Brockman J ed. (2014) *The Universe: Leading Scientists Explore the Origin, Mysteries and Future of the Cosmos*. Harper Perennial, New York

Brockman J ed. (2014) *What Should We Be Worried About? Real Scenarios that Keep Scientists Up at Night*. Harper Perennial, New York

Brockman J ed. (2015) *This Idea Must Die: Scientific Theories that are Blocking Progress*. Harper Perennial, New York

Brockman J ed. (2015) *What to Think About Machines That Think: Today's Leading Thinkers on the Age of Machine Intelligence*. Harper Perennial, New York

Brockman J ed. (2016) *Know This: Today's Most Interesting and Important Scientific Ideas, Discoveries, and Developments*. Harper Perennial, New York

Brockman J ed. (2017) *Life: The Leading Edge of Evolutionary Biology, Genetics, Anthropology, and Environmental Science*. Harper Perennial, New York

Brockman J ed. (2018) *This Idea is Brilliant: Lost, Overlooked, and Underappreciated Scientific Concepts Everyone Should Know*. Harper Perennial, New York
Brockman M ed. (2009) *What's Next? Dispatches on the Future of Science*. Vintage Books, New York
Bronowski J (1951) *The Common Sense of Science*. Faber and Faber, London
Bronowski J (1973) *The Ascent of Man*. BBC Books
Brooks M (2013) *Free Radicals: The Secret Anarchy of Science*. The Overlook Press, New York
Brooks M (2016) *At the Edge of Uncertainty: 11 Discoveries Taking Science by Surprise*. Overlook Press, New York
Brown P (1989) *The World of Late Antiquity: AD 150-750*. W.W. Norton, New York
Bryson B (2003) *A Short History of Nearly Everything*. Doubleday, London
Bryson B ed. (2010) *Seeing Further: The Story of Science & The Royal Society*. HarperCollins Publ., London, U.K.
Butterfield H (1957) *The Origins of Modern Science: 1300-1800*. The Free Press, New York
Bynum W (2008) *The History of Medicine: A Very Short Introduction*. Oxford Univ. Press
Bynum W (2012) *A Little History of Science*. Yale Univ. Press
Bynum W and Bynum H eds. (2011) *Great Discoveries in Medicine*. Thames & Hudson, London
Capra F (1975) *The Tao of Physics: An Exploration of the Parallels between Modern Physics and Eastern Mysticism*. Shambhala Publs., Boston, USA
Carey N (2012) *The Epigenetics Revolution: How Modern Biology is rewriting our Understanding of Genetics, Disease, and Inheritance*. Columbia Univ. Press, New York
Carey N (2015) *Junk DNA: A Journey Through the Dark Matter of the Genome*. Columbia Univ. Press, New York
Carroll S (2016) *The Big Picture: On the Origins of Life, Meaning, and the Universe Itself*. Dutton, New York
Chalmers A (2013) *What is this Thing called Science?* Univ. of Queensland Press, Australia
Chiras D (2016) *Environmental Science, 10th Ed*. Jones & Bartlett Learning, Burlington, MA, USA
Church G, Regis E (2012) *Regenesis: How Synthetic Biology will reinvent Nature and Ourselves*. Basic books, New York
Clegg B (2014) *The Quantum Age: How the Physics of the Very Small has Transformed out Lives*. Icon Books, London
Clegg B (2015) *Ten Billion Tomorrows: How Science Fiction Technology Became Reality and Shapes the Future*. St. Martin's Press, New York
Clegg B (2016) *Are Numbers Real? The Uncanny Relationship of Mathematics and the Physical World*. St. Martin's Press, New York
Close F (2009) *Nothing: A Very Short Introduction*. Oxford Univ. Press, Oxford, UK

Close F, Marten M, Sutton C (2002) *The Particle Odyssey: A Journey to the Heart of Matter*. Oxford Univ. Press, Oxford

Coles P (2001) *Cosmology: A Very Short Introduction*. Oxford Univ. Press

Coyne JA (2015) *Faith vs. Fact: Why Science and Religion are Incompatible*. Viking Penguin, New York

Craig E (2002) *Philosophy: A Very Short Introduction*. Oxford Univ. Press, Oxford

Crump T (2002) *A Brief History of Science as seen through the Development of Scientific Instruments*. Robinson, London, UK

Curd M, Cover JA, Pincock C eds. (2013) *Philosophy of Science: The Central Issues*. W. W Norton & Co., New York

Dampier-Whetham W (1930) *A History of Science and its Relations with Philosophy and Religion*. MacMillan Co., New York

Dartnell L (2014) *The Knowledge: How to Rebuild Civilization in the Aftermath of a Cataclysm*. Penguin, New York

Dartnell L (2017) *Apocalypse* in *What's Next? Even Scientists Can't Predict the Future – Or Can They?* (ed. J. Al-Khalili; Profile Books)

Darwin C (1839) *The Voyage of the Beagle*. Henry Colburn, London

Darwin C (1859) *The Origin of Species by Means of Natural Selection*. John Murray, London

Darwin C (1871) *The Descent of Man, and Selection in Relation to Sex*. John Murray, London

Darwin C (1872) *The Expression of the Emotions in Man and Animals*. John Murray, London

Dawkins R (2006) *The God Delusion*. Bantam Press, London

Deamer D (2011) *First Life: Discovering the Connections between Stars, Cells, and How Life Began*. Univ. of California Press, Berkeley CA

Dehaene S (2014) *Consciousness and the Brain: Deciphering How the Brain Codes our Thoughts*. Penguin books, New York

Deutsch D (1998) *The Fabric of Reality*. Penguin Books, London

DeWitt R (2010) *Worldviews: An Introduction to the History and Philosophy of Science*. Wiley-Blackwell, Chichester, UK

De Waal F (2016) *Are We Smart Enough to Know How Smart Animals Are?* W.W. Norton & Co., New York

Diamond J (1999) *Guns, Germs and Steel: The Fates of Human Societies*. W.W. Norton & Co., New York

Diamond J (2002) *The Rise and Fall of the Third Chimpanzee: How our Animal Heritage affects the Way We Live*. Vintage Books, London

Diamond J (2005) *Collapse: How Societies Choose to Fail or Succeed*. Penguin, New York

Dixon T (2008) *Science and Religion: A Very Short Introduction*. Oxford Univ. Press, Oxford

Doudna JA, Sternberg SH (2017) *A Crack in Creation: Gene Editing and the Unthinkable Power to Control Evolution*. Houghton Mifflin Harcourt, New York

Dyson F (2006) *The Scientist as Rebel*. New York Review Books, NY
Dyson F (2015) *Dreams of Earth and Sky*. New York Review Books, NY
Eagleton T (2007) *The Meaning of Life: A Very Short Introduction*. Oxford Univ. Press, Oxford, UK
Ehrlich PR (2002) *Human Natures: Genes, Cultures, and the Human Prospect*. Penguin, 2002
Enger ED, Smith BF (2013) *Environmental Science: A Study of Interrelationships, 13th Ed.* McGraw-Hill, New York
Enriquez J, Gullans S (2015) *Evolving Ourselves: How Unnatural Selection and Nonrandom Mutation are Changing Life on Earth*. Portfolio/Penguin, New York
Everett D (2017) *How Language Began: The Story of Humanity's Greatest Invention*. Liveright Publ. Co., New York
Fara P (2009) *Science: A Four Thousand Year History*. Oxford Univ. Press, Oxford
Finkel E (2012) *The Genome Generation*. Melbourne Univ. Press, Carlton, Victoria, Australia
Fossel M (2015) *The Telomerase Revolution*. BenBella Books, Dallas
Freely J (2012) *Before Galileo: The Birth of Modern Science in Medieval Europe*. Overlook Duckworth, London, UK
Gamble C, Gowlett J, Dunbar R (2014) *Thinking Big: How the Evolution of Social Life Shaped the Human Mind*. Thames & Hudson, London
Gamow G (1961) *The Great Physicists from Galileo to Einstein*. Dover Publ., New York
Giles J (2005) *Internet Encyclopaedias go Head to Head*. Nature 438, 900
Godfrey-Smith P (2003) *Theory and Reality: An Introduction to the Philosophy of Science*. Univ. of Chicago Press, Chicago
Godfrey-Smith P (2014) *Philosophy of Biology*. Princeton Univ. press, Princeton, New Jersey
Goldstein R (2006) *Incompleteness: The Proof and Paradox of Kurt Gödel*. W.W. Norton, New York
Gottleib A (2010) *The Dream of Reason: A History of Western Philosophy from the Greeks to the Renaissance*. W. W. Norton & Co., New York
Gottleib A (2016) *The Dream of Enlightenment: The Rise of Modern Philosophy*. Liveright Publ. Corp., New York
Goudie A (2006) *The Human Impact on the Natural Environment: Past, Present and Future*. Blackwell Publ., Malden, Mass., USA
Grant E (1996) *The Foundations of Modern Science in the Middle Ages: Their Religious, Institutional and Intellectual Contexts*. Cambridge University Press, Cambridge UK
Greene B (2000) *The Elegant Universe: Superstrings, Hidden Dimensions, and the Quest for the Ultimate Theory*. Vintage Books, New York
Greene B (2011) *The Hidden Reality: Parallel Universes and the Deep Laws of the Cosmos*. Vintage Books, New York
Gregory A (2003) *Eureka! The Birth of Science*. Icon Books, Cambridge UK
Gribbon J (2003) *Science: A History*. Penguin Books, London, UK

Gribbon J (2012) *Erwin Schrödinger and the Quantum Revolution.* Bantam Press, London

Gutfreund H, Renn J (2015) *The Road to Relativity: The History and Meaning of Einstein's 'The Foundation of General Relativity'.* Princeton Univ. Press, Princeton, NJ

Hanbury Brown R (2002) *There are no Dinosaurs in the Bible.* Chalkcroft Press, Penton Mewsey, UK

Harari YN (2014) *Sapiens: A Brief History of Humankind.* Signal Books (McClelland & Stewart), Canada

Harari YN (2015) *Homo Deus: A Brief History of Tomorrow.* Signal Books (McClelland & Stewart), Canada

Harris S (2010) *The Moral Landscape: How Science can Determine Human Values.* Free Press, New York

Harwit M (1981) *Cosmic Discovery: The Search, Scope and Heritage of Astonomy.* Basic Books, New York

Hawking S, Mlodinow L (2010) *The Grand Design: New Answers to the Ultimate Questions of Life.* Bantam Press, London

Heilbron JL (2015) *Physics: A Short History from Quintessence to Quarks.* Oxford Univ. Press, Oxford

Henry J (2008) *The Scientific Revolution and the Origins of Modern Science.* Palgrave Macmillan, UK

Holloway R (2016) *A Little History of Religion.* Yale Univ. Press, New Haven, Conn., USA

Homer-Dixon T (2006) *The Upside of Down: Catastrophe, Creativity and the Renewal of Society.* Island Press, Washington, DC

Horgan J (1996) *The End of Science: Facing the Limits of Knowledge in the Twilight of the Scientific Age.* Addison Wesley

Hoskin M (2003) *The History of Astronomy: A Very Short Introduction.* Oxford Univ. Press, Oxford, UK

Huff TE (2011) *Intellectual Curiosity and the Scientific Revolution: A Global Perspective.* Cambridge Univ. Press, Cambridge, UK

Huff TE (2017) *The Rise of Early Modern Science: Islam, China, and the West.* Cambridge Univ. Press, Cambridge, UK

Isaacson W (2008) *Einstein: His Life and Universe.* Simon & Schuster, New York

James CR (2014) *Science Unshackled: How Obscure, Abstract, Seemingly Useless Scientific Research Turned Out to be the Basis for Modern Life.* Johns Hopkins Univ. Press, Baltimore, USA

Jastrow R, Rampino M (2008) *Origins of Life in the Universe.* Cambridge Univ. Press, Cambridge, UK

Johnson G (2004) *A Shortcut Through Time: The Path to the Quantum Computer.* Vintage Books, London

Jones BF, Reedy EJ, Weinberg BA (2014) *Age and Scientific Genius.* In *The Wiley Handbook of Genius* (ed. Simonton DK), pp. 422-450, Wiley-Blackwell

Kaku M (2012) *Physics of the Future: How Science will shape Human Destiny and our Daily Lives by the year 2100*. Anchor Books, New York

Kaku M (2014) *The Future of the Mind: The Scientific Quest to Understand, Enhance, and Empower the Mind*. Doubleday, New York

Kellermann K, Sheets B (1983) *Serendipitous Discoveries in Radio Astronomy*. The National Radio Astronomy Observatory, Green Bank, West Virginia, USA

Krauss LM (2012) *A Universe from Nothing: Why There is Something Rather than Nothing*. Free Press, New York

Krauss LM (2017) *The Greatest Story Ever Told – So Far: Why are We Here?* Atria Books, New York

Kuhn TS (1962) *The Structure of Scientific Revolutions*. The Univ. of Chicago Press, Chicago

Kurzweil R (2006) *The Singularity is Near: When Humans Transcend Biology*. Penguin, New York

Lane N (2015) *The Vital Question: Energy, Evolution, and the Origins of Complex Life*. W.W. Norton & Co., New York

Lanza R, Berman B (2016) *Beyond Biocentrism: Rethinking Time, Space, Consciousness, and the Illusion of Death*. BenBella Books, Dallas, Texas

Larsen PO, von Ins M (2010) *The Rate of Growth in Scientific Publication and the Decline in Coverage Provided by Science Citation Index*. Scientometrics Vol. 84, p. 575

Larson EJ, Witham L (1998) *Leading scientists still reject God*. Nature 394, 313

Laughlin RB (2005) *A Different Universe: Reinventing Physics from the Bottom Down*. Basic Books, New York

Levitin DJ (2007) *This is Your Brain on Music: The Science of a Human Obsession*. Plume Books, New York

Lindberg DC ed. (1978) *Science in the Middle Ages*. The Univ. of Chicago Press, Chicago and London

Lindberg DC (1992) *The Beginnings of Western Science: The European Scientific Tradition in Philosophical, Religious, and Institutional Context, Prehistory to A.D. 1450*. The Univ. of Chicago Press, Chicago and London

Lindley D (2007) *Uncertainty: Einstein, Heisenberg, Bohr, and the Struggle for the Soul of Science*. Doubleday, New York

Livio M (2005) *The Equation That Couldn't Be Solved: How Mathematical Genius Discovered the Language of Symmetry*. Simon & Schuster, New York

Livio M (2009) *Is God A Mathematician?* Simon & Schuster, New York

Livio M (2013) *Brilliant Blunders: From Darwin to Einstein – Colossal Mistakes by Great Scientists that Changed our Understanding of Life and the Universe*. Simon & Schuster, New York

Livio M (2017) *Why? What Makes Us Curious*. Simon & Schuster, New York

Maddox J (1999) *What Remains to be Discovered? Mapping the Secrets of the Universe, The Origins of Life, and the Future of the Human Race*. Touchstone, New York

McClellan J III, Dorn H (2006) *Science and Technology in World History, An Introduction. 2nd Ed.* Johns Hopkins Univ. Press
McFadden J, Al-Khalili J (2014) *Life on the Edge: The Coming of Age of Quantum Biology.* Broadway Books, New York
Merali Z (2017) *A Big Bang in a Little Room: The Quest to Create New Universes.* Basic Books, New York
Mesler B, Cleaves HJ (2016) *A Brief History of Creation: Science and the Search for the Origin of Life.* W.W. Norton & Co., New York
Meyers MA (2011) *Happy Accidents: Serendipity in Major Medical Breakthroughs in the Twentieth Century.* Arcade Publishing, New York
Mlodinow L (2013) *Subliminal: How your Unconscious Mind Rules your Behavior.* Vintage Books, New York
Mlodinow L (2015) *The Upright Thinkers: The Human Journey from Living in Trees to Understanding the Cosmos.* Pantheon Books. New York
Morris R (2002) *The Big Questions: Probing the Promise and Limits of Science.* Times Books, New York
Motesharrei S, Rivas J, Kalnay E (2014) *Human and Nature Dynamics. Ecological Economics* vol. 101, 90-102
Mukherjee S (2016) *The Gene: An Intimate History.* Scribner, New York
Nagel T (2012) *Mind & Cosmos: Why the Materialist Neo-Darwinian Conception of Nature is almost Certainly False.* Oxford Univ. Press
Narison S (2016) *Particles and the Universe: From the Ionian School to the Higgs Boson and Beyond.* World Scientific, Singapore
Newton I (1687) *Philosophiae Naturalis Principia Mathematica.* S. Pepys, London
Nicolaides D (2014) *In the Light of Science: Our Ancient Quest for Knowledge and the Measure of Modern Physics.* Prometheus Books, Amherst, New York
Nixey C (2017) *The Darkening Age: The Christian Destruction of the Classical World.* Macmillan, London
Ocampo et al. (2016) *In Vivo Amelioration of Age-Associated Hallmarks by Partial Reprogramming.* Cell 167, 1719–1733
Oerter R (2006) *The Theory of Almost Everything: The Standard Model, the Unsung Triumph of Modern Physics.* Plume, New York
Okasha S (2002) *Philosophy of Science: A Very Short Introduction.* Oxford Univ. Press, Oxford
Otto S (2016) *The War on Science: Who's Waging It, Why It Matters, What We Can Do about It.* Milkweed Editions, Minneapolis, U.S.
Penrose R (2016) *Fashion Faith and Fantasy in the New Physics of the Universe.* Princeton Univ. Press, Princeton
Pinker S (2012) *The Better Angels of Our Nature: A History of Violence and Humanity.* Penguin Books, London
Pinker S (2018) *Enlightenment Now: The Case for Reason, Science, Humanism and Progress.* Viking, New York

Popper K (2002) *The Logic of Scientific Discovery*. Routledge Classics, London and New York

Price DJ de Solla (1961) *Science since Babylon*. Yale Univ. Press, New Haven CT

Price DJ de Solla (1963) *Little Science. Big Science*. Columbia Univ. Press, NY

Principe LM (2011) *The Scientific Revolution: A Very Short Introduction*. Oxford Univ. Press, Oxford

Randall L (2013) *Higgs Discovery: The Power of Empty Space*. HarperCollins, New York

Randers J (2012) *2052: A Global Forecast for the Next Forty Years*. Chelsea Green Publishing, White River Junction, Vermont

Read L (2012) *I, Pencil: My Family Tree as Told to Leonard E. Read*. Foundation for Economic Education

Rees M (2003) *Our Final Century: Will the Human Race Survive the Twenty-first Century?* William Heinemann, London

Rees M (2011) *From Here to Infinity: Scientific Horizons*. Profile Books, London

Ridley M (2015) *The Evolution of Everything: How New Ideas Emerge*. HarperCollins Publ., New York

Ripple et al. (2017), *World Scientists' Warning to Humanity: A Second Notice*. Bioscience Volume 67, Issue 12, p. 1026

Roberts RM (1989) *Serendipity: Accidental Discoveries in Science*. John Wiley & Sons, Inc., New Jersey, USA

Robinson A ed. (2012) *The Scientists: An Epic of Discovery*. Thames & Hudson, London

Rovelli C (2017) *Reality Is Not What It Seems: The Journey to Quantum Gravity*. Riverhead Books, New York

Russell B (1960) *A History of Western Philosophy*. Simon & Schuster, New York

Serafini A (1993) *The Epic History of Biology*. Perseus Publishing, Cambridge, MA

Seung S (2013) *Connectome: How the Brain's Wiring Makes Us Who We Are*. Mariner Books, New York

Shaver P (2011) *Cosmic Heritage: Evolution from the Big Bang to Conscious Life*. Springer, Heidelberg

Shlain L (2014) *Leonardo's Brain: Understanding da Vinci's Creative Genius*. LP, Guilford, Connecticut, USA

Shubin N (2009) *Your Inner Fish: A Journey into the 3.5-Billion-Year History of the Human Body*. Vintage Books, New York

Smolin L (2007) *The Trouble with Physics: The Rise of String Theory, the Fall of a Science, and What Comes Next*. Mariner Books, New York

Smolin L (2013) *Time Reborn: From the Crisis in Physics to the Future of the Universe*. Vintage Canada, Toronto

Stirrat M, Cornwall RE (2013) *Eminent scientists reject the supernatural: a survey of the Fellows of the Royal Society*. Evolution, Education and Outreach 20136, 33

Stringer C (2012) *Lone Survivors: How We Came to be the Only Humans on Earth*. St. Martin's Griffin, New York

Suddendorf T (2013) *The Gap: The Science of What Separates us from other Animals.* Basic Books, New York

Sykes B (2001) *The Seven Daughters of Eve: The Science that Reveals our Genetic History*. W.W. Norton & Co., New York

Tainter JA (1990) *The Collapse of Complex Societies (New Studies in Archaeology).* Cambridge Univ. Press

Taylor FS (1949) *A Short History of Science & Scientific Thought*. W.W. Norton & Co, New York

Tegmark M (2017) *Life 3.0: Being Human in the Age of Artificial Intelligence*. Alfred A. Knopf, New York

Trumble DR (2013) *The Way of Science: Finding Truth and Meaning in a Scientific Worldview*. Prometheus Books, Amherst, New York

Turok N (2013) *The Universe Within: From Quantum to Cosmos*. Allen & Unwin, Sydney, Australia

Venter JC (2013) *Life at the Speed of Light: From the Double Helix to the Dawn of Digital Life*. Viking Penguin, New York

Wade N (2006) *Before the Dawn: Recovering the Lost History of our Ancestors.* Penguin Books, New York

Wade N (2014) *A Troublesome Inheritance: Genes, Race and Human History*. Penguin Press, New York

Walter C (2013) *Last Ape Standing: The Seven-Million-Year Story of How and Why We Survived.* Bloomsbury, New York

Ward K (2008) *The Big Questions in Science and Religion*. Templeton Foundation Press, West Conshohocken, Pennsylvania, USA

Weatherall J (2016) *Void: The Strange Physics of Nothing*. Yale Univ. Press, New Haven, Conn., USA

Weinberg S (1992) *Dreams of a Final Theory: The Scientist's Search for the Ultimate Laws of Nature*. Pantheon, New York

Weinberg S (2009) *Lake Views: This World and the Universe*. Harvard Univ. Press, Cambridge, USA

Weinberg S (2015) *To Explain the World: The Discovery of Modern Science.* HarperCollins, New York

Wiggins AW, Wynn CM (2016) *The Human Side of Science: Edison and Tesla, Watson and Crick, and other Personal Stories Behind Science's Big Ideas*. Prometheus books, Amherst, NY

Wilczek K (2008) *The Lightness of Being: Mass, Ether, and the Unification of Forces.* Basic Books, New York

Wilczek K (2015) *A Beautiful Question: Finding Nature's Deep Design*. Penguin Press, New York

Wilson EO (1998) *Consilience: The Unity of Knowledge*. Alfred A. Knopf, New York

Wilson EO (2012) *The Social Conquest of Earth*. Liveright Publ. Co., New York

Wilson EO (2014) *The Meaning of Human Existence*. Liveright Publ. Co., New York

Winters RW (2016) *Accidental Medical Discoveries: How Tenacity and Pure Dumb Luck Changed the World*. Skyhorse Publ., New York

Wiseman R (2015) *Paranormality: The Science of the Supernatural*. Pan Books, London

Wootton D (2015) *The Invention of Science: A New History of the Scientific Revolution*. HarperCollins Publ., New York

Wulf A (2013) *Chasing Venus: The Race to Measure the Heavens*. Vintage Books, New York

Yanofsky NS (2013) *The Outer Limits of Reason: What Science, Mathematics and Logic Cannot Tell Us*. MIT Press, Cambridge, MA

Subject Index

P. Shaver, *The Rise of Science*, https://doi.org/10.1007/978-3-319-91812-9

Name Index

P. Shaver, *The Rise of Science*, https://doi.org/10.1007/978-3-319-91812-9

Printed in the United States
By Bookmasters